Lithium-ion and Lithium–Sulfur Batteries

Fundamentals to performance

Online at: https://doi.org/10.1088/978-0-7503-4881-2

Lithium-ion and Lithium–Sulfur Batteries

Fundamentals to performance

Sandeep A Arote
Sangamner Nagarpalika Arts, D. J. Malpani Commerce and B. N. Sarda Science College (Autonomous) Sangamner, affiliated to Savitribai Phule Pune University, M.S. India

IOP Publishing, Bristol, UK

ISBN 978-0-7503-4881-2 (ebook)
ISBN 978-0-7503-4879-9 (print)
ISBN 978-0-7503-4882-9 (myPrint)
ISBN 978-0-7503-4880-5 (mobi)

DOI 10.1088/978-0-7503-4881-2

Version: 20221201

IOP ebooks

British Library Cataloguing-in-Publication Data: A catalogue record for this book is available from the British Library.

Published by IOP Publishing, wholly owned by The Institute of Physics, London

IOP Publishing, No.2 The Distillery, Glassfields, Avon Street, Bristol, BS2 0GR, UK

US Office: IOP Publishing, Inc., 190 North Independence Mall West, Suite 601, Philadelphia, PA 19106, USA

Contents

Preface

The content of this book, *Lithium-ion and Lithium–sulfur Batteries: Fundamentals to Performance*, focuses on energy storage technologies, namely lithium-ion and lithium–sulfur batteries. It will acquaint readers with the fundamentals of secondary batteries, working mechanisms, electrode materials, challenges, and opportunities for further development in lithium-ion battery (LIB) technology. Massive efforts have been made over the last 20 years to design and develop electrode materials with excellent electrical conductivity, high specific surface area, high theoretical capacity, exceptional thermal, chemical, and structural stability, and low cost. High energy density batteries for heavy electric vehicle applications can be developed by a comprehensive understanding of the batteries' working mechanism, redox kinetics, and the properties of the electrode materials. In line with this, the content of this book is divided into four chapters. The first chapter presents an overview of the key concepts, brief history of the advancement in battery technology, and the factors governing the electrochemical performance metrics of battery technology. It also includes in-depth explanations of electrochemistry and the basic operation of lithium-ion batteries. The second chapter provides a detailed overview of the numerous cathode materials that have been extensively researched along with their advantages and disadvantages. It also highlights the electrochemical reaction mechanisms involved in Li-ion battery operation. In order to gain additional improvements, modifications to adopted high-capacity cathode materials are being made; these modifications are described for each class of material. Lithium–sulfur batteries (LiSBs) have been considered to have the most potential as next-generation electrochemical energy storage systems due to their high theoretical energy density. LiSBs have the potential to accomplish around five to six times better performance than current LiBs in terms of specific capacity and energy density. The fundamental aspects of LiSBs, working mechanisms, key requirements for each of the components, and SWOT analysis of LiSBs are covered in detail in chapter three. The fourth chapter discusses the performance of LiSBs with different sulfur host materials, with a focus on sulfur hosts based on carbon and its many derivatives, metal oxides, and metal sulfides. The efforts made to enhance the performance of materials used frequently for electrodes in LIBs and LiSBs have been discussed extensively. The electrochemical performance of each different class of material is summarized in a tabulated format whenever suitable. Every effort has been made to include topics that will be beneficial to readers who are studying graduate, postgraduate, and research degrees in the field of energy storage devices as well as to give in-depth knowledge to a person pursuing a technical profession in R&D for the battery industry.

Sandeep A Arote

Acknowledgments

I would like to begin by thanking my angels Shrinidhi and Nirja for their support and encouragement while I have undertaken this project. Without your patience and support over the many weekends and evenings, this book would not have been possible. Reejul, Shaili, Sharvil, Akka, Dada, Darshana, Sheetal, and Sonali deserve a particular thank you as well. They constantly tell me that 'there is plenty of room at the bottom.'

The suggestions and final editing of the book were greatly helped by Mr Abhijit Landge, Mr Yogesh Hase, Mr Abbas Pathan, and Mr Avadhut Landge. They provided me unconditional support during the entire project, both physically and intellectually. I am very grateful to them.

It is my pleasure to express my due thanks to Dr Sanjay Malpani, Chairman, and all the office bearers of the Shikshan Prasarak Sanstha, Prof. Rangnath Pathare, Prof. Dr Arun Gaikwad, Principal, Dr Sanjaykumar Dalvi, HOD (Physics) Sangamner College, Sangamner, for their continual and invaluable support and advice.

I would like to extend a very special thanks to Principal T N Kanawade, Prof. Pradip Ghule, Prof. S P Landge, Dr Ravindra Tasildar, Dr Anirudhha Mandlik, and Dr Maruti Kusmude for their continued support during completion of this work. I also want to acknowledge all the students, colleagues, and friends of Sangamner College who have shared their views and enthusiasm with me. They have been motivating.

I am also grateful to the publishers, authors, and other copyright owners who gave consent for me to utilize their figures to enhance the quality of the present work.

Last, but definitely not least, I want to express my gratitude to the entire IOP team for their invaluable help and prompt support. Ashley Gasque Stec, Senior Commissioning Editor, IOP Publishing, deserves special recognition for her unwavering support to this project from the very beginning and for making it possible to put all the difficult puzzle pieces together.

Sandeep Annasaheb Arote

Author biography

Sandeep A Arote

Dr Sandeep A Arote is an Associate Professor in Physics at the Sangamner Nagarpalika Arts, D. J. Malpani Commerce and B. N. Sarda Science College, (Autonomous) Sangamner, affiliated to Savitribai Phule Pune University, (Former University of Pune), MS, India. He has 16 years' teaching and 11 years' research experience. He has published over 40 research articles in reputed peer-reviewed international journals and also presented his research work in several national and international conferences. His primary research interests are in nanostructured materials for sensors and energy generating and storage technologies such as solar cells, batteries, and supercapacitors. He is a co-author of several undergraduate physics textbooks, including 'Handbook of Physics, 'Practical Course in Physics', 'Solid State Physics', 'Nuclear Physics', and a booklet on graph plotting methods. Recently, he has published a book entitled *Electrochemical Energy Storage Devices and Supercapacitors: An overview* (2021, IOP Publishing Ltd, Bristol, UK).

Abbreviations

1D	One-dimensional
2D	Two-dimensional
3D	Three-dimensional
CB	Carbon black
CEI	Cathode electrolyte interphase
CMK	Carbon mesostructured by KAIST
CNFs	Carbon nanofibers
CNS	Carbon nanosheets
CNTs	Carbon nanotubes
CV	Cyclic voltammetry
CS	Carbon sphere
CVD	Chemical vapor deposition
DEC	Diethyl carbonate
DMC	Dimethyl carbonate
DOD	Depth of discharge
EV	Electric vehicle
EC	Ethylene carbonate
ECC	Electrochemical cell
EMC	Ethyl methyl carbonate
ESD	Electrostatic or electro spinning spray deposition
G	Graphene
GO	Graphene oxide
GNS	Graphene nanosheets
GCD	Galvanostatic charge discharge
GITT	Galvanostatic intermittent titration technique
HCS	Hollow/Hierarchical carbon spheres
HOMO	Highest occupied molecular orbital
IEA	International Energy Association
LBS	Lithium borosilicate
LCO	Lithium cobalt oxide
LFP	Lithium iron phosphate
LIB	Lithium-ion battery
LiSB	Lithium–sulfur battery
LMO	Lithium manganese oxide
LNO	Lithium nickel oxide
LPSs	Lithium polysulfides
LUMO	Lowest unoccupied molecular orbital
MCHS	Mesoporous carbon hollow spheres
NGCNs	N-doped graphitic carbon nanoshells
MP	Melamine phosphate
MOFs	Metal–organic frameworks
NCA	Nickel cobalt aluminum oxide
NMC	Nickel manganese cobalt
OCV	Open-circuit voltage
PAA	Polyacrylic acid
PAN	Polyacrylonitrile
PBs	Primary batteries

PC	Propylene carbonate
PEO	Polyethylene oxide
PMMA	Poly methyl methacrylate
PSs	Polysulfides
PTC	Positive temperature coefficient
PTFE	Polyetrafluoroethylene
PVA	Polyvinyl alcohol
PVDF	Polyvinylidene fluoride
QDs	Quantum dots
rGO	Reduced graphene oxide
RF	Radio frequency
SBs	Secondary batteries
SEI	Solid electrolyte interphase
SEM	Scanning electron microscopy
SHE	Standard hydrogen electrode
SOC	State of charge
STEM-DF	Scanning transmission electron microscopy-dark-field
TEM	Transmission electron microscopy
TMS	Transition metal sulfide
UMC	Ultra microporous carbon
XPS	X-ray photoelectron spectroscopy

Chemical element/compounds and their name/formulae

Ag	Silver
Al	Aluminum
Al_2O_3	Aluminum oxide
B	Boron
C	Carbon
CaO	Calcium oxide
$CaCO_3$	Calcium carbonate
CeO_2	Cerium dioxide
Cd	Cadmium
Co	Cobalt
CO_2	Carbon dioxide
CoO	Cobalt (II) oxide (CoO_2, CoO_6)
Co_3O_4	Cobalt tetraoxide
Co_2S_3	Cobalt sulfide (Co_3S_4, Co_4S_3 and Co_9S_8)
Co_2P	Cobalt phosphide
Co_3S_4@S	Cobalt sulfide–sulfur nanocomposites
Cr	Chromium
Cu	Copper
Fe	Iron
Fe_2O_3	Ferric oxide
Fe_2P	Iron phosphide
Fe_3C	Iron carbide
FeS_2	Iron (II) disulfide
Ga	Gallium
Gd	Gadolinium

K_2CO_3	Potassium carbonate
La_2O_3	Lanthanum (III) oxide
Li	Lithium
$LiAsF_6$	Lithium hexafluoroarsenate
$LiBF_4$	Lithium tetrafluoroborate
$LiClO_4$	Lithium perchlorate
$LiCoO_2$	Lithium cobalt oxide
Li_2CO_3	Lithium carbonate
$LiCoPO_4$	Lithium cobalt phosphate
LiF	Lithium fluorine
$LiFeBO_3$	Lithium iron borate
$LiFePO_4$	Lithium iron phosphate
$LiMO_2$	Lithium layered oxides (M = Mn, Ni, Co, Fe, Cr, etc)
$LiMnPO_4$	Lithium manganese (II) phosphate
$LiMnO_2$	Lithium manganese dioxide
$LiMn_2O_4$	Lithium manganese oxide
$LiMnO_4$	Lithium permanganate
$LiMnCoO_4$	Lithium manganese cobalt oxide
$LiMnCrO_4$	Lithium manganese chromium oxide
$LiNiO_2$	Lithium nickel oxide
$LiNiMnCoO_2$	Lithium nickel/manganese/cobalt (NMC)
$LiNiPO_4$	Lithium nickel phosphate
LiO_2	Lithium oxide
LiO_6	Lithium oxide octahedral
Li_2O	Lithium oxide
LiOH	Lithium hydroxide
$LiPF_6$	Lithium hexafluorophosphate
Li-S	Lithium sulfide
Li_2SO_4	Lithium sulfate
$Li_3V_2(PO_4)_3$	Lithium vanadium phosphate
$LiOH{\cdot}H_2O$	Lithium hydroxide
$LiAlSiO_4$	Lithium aluminum orthosilicate
Li_2FeSiO_4	Lithium iron orthosilicate
Li_2MnSiO_4	Lithium manganese orthosilicate
Li_2CoSiO_4	Lithium cobalt orthosilicate
Li_2NiSiO_4	Lithium nickel orthosilicate
$LiMnBO_3$	Lithium manganese borate
$LiCoBO_3$	Lithium cobalt borate
$LiNiBO_3$	Lithium nickel borate
$Li_4Ti_5O_{12}$	Lithium titanate
LiV_2O_5	Lithium vanadium pentoxide
LiV_3O_8	Lithium trivanadate
$LiNiVO_4$	Lithium nickel vanadate
$LiCoVO_4$	Lithium cobalt vanadate
Mg	Magnesium
MgO	Magnesium oxide
$Mg_{0.6}Ni_{0.4}O$	Magnesium nickel oxide
Mn	Manganese
MnO_2	Manganese dioxide
Mn_3O_4	Manganese tetroxide

Mn_3O_4@CNF@S	Trimanganese tetroxide-carbon nanofibers–sulfur nanocomposite
Mo	Molybdenum
MoO_2	Molybdenum dioxide
MoO_3	Molybdenum trioxide
Mo_2C	Molybdenum carbide
MoO_3@CNT	Molybdenum trioxide-carbon nanotube nanocomposite
MoS_2	Molybdenum disulfide
MoS_2@ HCS	Molybdenum disulfide-hierarchical carbon spheres nanocomposite
N	Nitrogen
Na	Sodium
Na_2CO_3	Sodium carbonate
Nb	Niobium
NbS_2	Niobium (IV) sulfide
Nb_2O_5	Niobium (V) oxide
Ni	Nickel
Ni–Cd	Nickel–cadmium
$NiCo_2O_4$	Nickel cobaltite (Nickel cobalt oxide)
Ni-Fe	Nickel-iron
$NiFe_2O_4$	Nickel iron oxide (Nickel ferrite)
NMP	N-methyl pyrrolidone
NiMH	Nickel metal hydride
NiO	Nickel oxide
Ni_2P	Nickel phosphide
NiOOH	Nickel oxide hydroxide
$Ni(OH)_2$	Nickel (II) hydroxide
Ni–Zn	Nickel–zinc
O	Oxygen
o-$LiMnO_2$	Orthorhombic-lithium manganese dioxide
PEDOT	Poly(3,4-ethylenedioxythiophene)
PO_4^{3-}	Phosphate anion
Ru	Ruthenium
rGO@VS_2	Reduced graphene oxide-vanadium disulfide nanocomposite
$ROCO_2Li$	Lithium alkyl carbonates
S	Sulfur
Sc	Scandium
S_8	Octasulfur
SiO_2	Silicon dioxide
Sm	Samarium
Sn	Tin
SnO_2	Tin (IV) oxide
$S_2O_3^{2-}$	Thiosulfate polythionate (thiosulfate ion)
S@CB	Sulfur-carbon black nanocomposite
S@MWCNT	Sulfur-multiwall carbon nanotube nanocomposite
Ta	Tantalum
Tb	Terbium
Ti	Titanium
TiC	Titanium carbide
TiN	Titanium nitride
(TiO, TiO_2, Ti_4O_7)	Titanium oxide
TiS_2	Titanium disulfide

V	Vanadium
VO_2	Vanadium dioxide
V_2O_3	Vanadium trioxide
V_2O_5	Vanadium pentoxide
VN	Vanadium nitride
VS_2	Vanadium disulfide
WS_2	Tungsten (IV) disulfide
Zn	Zinc
ZnO	Zinc oxide
Zr	Zirconium
ZrO_2	Zirconium dioxide

Symbols

A	Ampere
Å	Angstrom
C	C-rate
°C	Degree Celsius
D	Atomic weight of the element
F	Faradays constant
G	Gibbs free energy
H	Enthalpy
I	Current
K	Kelvin
N	Total number of electrons
T	Temperature
S	Entropy
P	Pressure
V	Volume
V	Voltage
ρ	Electrical resistivity
σ	Electrical conductivity
μ	Electrochemical potential
μ_A	Chemical potential of anode
μ_C	Chemical potential of cathode
R_{int}	Internal resistance
E_{int}	Total internal energy
ΔS	Change in entropy
ΔV	Change in volume
ΔU	Change in internal energy
ΔG	Change in Gibbs free energy
Wt	Weight

Units

A	Ampere
h	Hour
Ah	Ampere hour
C	Coulomb
s	Second
m	Meter
m^2	Square meter
V	Volt
Pa	Pascal
TPa	Terapascal
nm	Nanometer
amu	Atomic mass unit
Ah g^{-1}	Ampere hour per gram
cm^2 s^{-1}	Square centimeter per second
cm^3 g^{-1}	Cubic centimeter per gram
g mol^{-1}	Gram per mole
g cm^{-3}	Gram per cubic centimeter
g Ah^{-1}	Gram ampere per hour
g cm^{-3}	Gram per cubic centimeter
GW	Giga watt
mA	Milliampere
mAh	Milliampere hour
mAh g^{-1}	Milliampere hour per gram
mAh cm^{-3}	Milliampere hour per cubic centimeter
mAh cm^{-2}	Milliampere hour per square centimeter
mA cm^{-2}	Milliampere per square centimeter
mg cm^{-2}	Milligram per square centimeter
m^2 g^{-1}	Square meter per gram
mV s^{-1}	Millivolt per second
nm^3	cubic nanometer
KJ mol^{-1}	Kilojoule per mole
J kg^{-1} K^{-1}	Joule per kilogram kelvin
Ω m	Ohm meter
S m^{-1}	Siemens per meter
S cm^{-1}	Siemens per centimeter
μm m^{-1} K^{-1}	Micrometer per meter kelvin
Wh kg^{-1}	Watt-hour per kilogram
W $m^{-1}k^{-1}$	Watt per meter kelvin
Wh L^{-1}	Watt-hour per liter

Chapter 1

Fundamentals and perspectives of lithium-ion batteries

Recently, energy conversion and storage have assumed a prominent role in the global growth of science and technology. Electric energy has become crucial for everything from portable consumer devices to electric hybrid vehicles. The invention of an energy storage system with high energy and power density could be the answer to the problems of the energy crisis and environmental degradation. In order to overcome these challenges, new battery chemistries are being researched as alternatives to conventional ones. One of the modern energy storage technologies with the highest commercial demand is lithium-ion batteries. They have a wide range of applications, from portable electronics to electric vehicles. Because of their light weight and high energy density, they are economically viable. This chapter presents an overview of the key concepts, a brief history of the advancement and factors governing the electrochemical performance metrics of battery technology. It also contains in-depth explanation of the electrochemistry and basic operation of lithium-ion batteries. An overview of LIB types and their manufacturing process is also provided. Consideration has also been given to the best anodes, cathodes, and electrolytes for Li-ion batteries in light of recent developments in the materials used to make those components.

1.1 Introduction

Over the past few decades, the world's industries and population have grown quickly, which has unexpectedly boosted the demand for energy. The heavy reliance on conventional energy sources like coal and crude oil, which are continuously decreasing and have led to a multitude of environmental and social problems, highlights the need for a sustainable, clean, and abundant energy source. Today, efficient energy conversion from renewable energy sources is one of the key industries to meet the world's primary energy consumption needs. The development of energy storage technology is seen in tandem with the expanding share of

doi:10.1088/978-0-7503-4881-2ch1

renewable energy in global energy production. Battery technology is constantly improving, allowing for effective and inexpensive energy storage. A battery is a common device of energy storage that uses a chemical reaction to transform chemical energy into electric energy. In other words, the chemical energy that has been stored is converted into electrical energy. A battery is composed of tiny individual electrochemical units, often known as electrochemical cells (ECCs). Any ECC consists of three basic components: anode, cathode, and electrolyte. For energy utilization the terminals of the cell are connected via an external circuit. Due to a charge imbalance, electrons move through the external circuit and, simultaneously, electrolyte ions move inside the cell, opposite to the direction of the electrons [1, 2].

To understand how batteries have changed through time and the potential for continued growth, it is vital to understand their basic functions, types, components, and performance criteria. The following sections in this chapter discuss the working mechanism of ECCs, the various types of batteries, battery components, fundamental terminologies, and important factors that will enable the development of a new battery technology.

1.2 Battery types and history [3–6]

Batteries are typically classified into two broad categories: primary batteries and secondary batteries, according to the structure of ECCs that they hold.

1.2.1 Primary batteries

Primary batteries (PBs) are single-use, non-rechargeable batteries as they store and give energy but cannot be recharged. They must be discarded after use since the chemical process that creates electricity while in use cannot be stopped. These batteries are frequently used in household items like radios, watches, remote controls, toys, and other items that don't require a lot of energy. These are produced in accordance with international standards and are affordable, secure, low maintenance, and practical to use.

Examples: dry cell and alkaline battery.

A dry cell need not be dry, rather it consists of an electrolyte in the form of paste. Although it can be used in either direction, the issue of electrolyte leakage is a significant barrier to long-term storage. Zinc–carbon batteries are the most common example. Alkaline batteries have more energy storage capacity and less electrolyte leakage than zinc–carbon batteries. They usually use potassium hydroxide, an alkaline electrolyte. They cost more than zinc–carbon batteries but perform better in extreme weather conditions. Examples include lithium–MnO_2, silver oxide, and alkaline–MnO_2.

1.2.2 Secondary batteries

Secondary batteries (SBs) are multi-use rechargeable batteries because they constantly store and supply energy over numerous charging and discharging cycles. Utilizing an external current, the chemical reaction that generates electricity can be reversed while in use. They are often used in portable consumer devices such inverters, telephones, computers, cameras, etc.

There are various types of SBs, depending on the electrolyte used and the electrodes' chemical composition. Both wet and dry cells can be SBs. The lead–acid battery, which uses electrodes of lead alloy and lead oxide as well as diluted sulfuric acid as the electrolyte, is the most common example of a wet cell with a liquid electrolyte. The lithium-ion battery used in computers and mobile devices is the most common illustration of a dry cell with electrolyte in the form of paste. The usage of SBs in hybrid electric vehicles is one of the fascinating new applications nowadays. Nickel–metal hydride (NiMH), nickel–cadmium (NiCd), and nickel–zinc (NiZn) batteries are some examples of SBs that are used often.

1.2.3 Historical milestones in the development of batteries

Around 1800, an Italian scientist, *Alessandro Volta*, developed the first 'real' battery, and demonstrated this using a pile of zinc and silver sheets with cloth soaked in salt water. In Volta's cell, the zinc acts as the anode and silver as the cathode. The electrons moved from the anode to cathode through the external circuit which connects them. The net electromotive force for electrons comes from redox reactions associated with the electrolyte–electrode interfaces. In 1836, a British chemist, *John Frederic Daniell*, invented the Daniell cell, which has copper as the cathode (positive electrode) immersed in copper (II) sulfate and zinc as the anode (negative electrode) immersed in zinc sulfate or dilute sulfuric acid solution. The two solutions are placed in a container, which is divided into two compartments, and are separated by a porous ion membrane partition. When electrodes are connected though the external circuit, the current starts to flow due to an imbalance in the chemical potential, causing reactions at the electrode surface as follows:

$$\text{Oxidation at anode: } Zn_{(s)} \rightarrow Zn^{2+}_{(aq)} + 2e^{-} \tag{1.1}$$

$$\text{Reduction at cathode: } Cu^{2+}_{(aq)} + 2e^{-} \rightarrow Cu_{(s)} \tag{1.2}$$

Thus, the whole reaction during current flow is:

$$Zn_{(s)} + Cu^{2+}_{(aq)} \rightarrow Zn^{2+}_{(aq)} + Cu_{(s)} \tag{1.3}$$

Thus, the Daniell cell is composed of two half cells of which neither works alone. After that at around 1859, a French physicist, *Gaston Planté*, invented the lead–acid battery, which was the first rechargeable battery to be commercially produced. It comprises lead alloy as the anode and lead oxide as the cathode immersed in dilute sulfuric acid electrolyte. During discharge, a chemical reaction occurs at both electrode surfaces as shown in equation (1.4), which leads to the formation of lead sulfate in the cell.

$$PbO_2 + 2H_2SO_4 + Pb \leftrightharpoons 2PbSO_4 + 2H_2O \tag{1.4}$$

This reaction can be reversed by passing an external current through the battery.

Later on, in 1866, a French electrical engineer, *Georges Leclanché*, invented the zinc–carbon battery, the first widely used battery in the world. The cell was also called the Leclanché cell, and consists of zinc as the anode, manganese dioxide with carbon as the cathode wrapped in a porous material, and ammonium chloride solution as the electrolyte. For commercial use the liquid electrolyte was replaced by paste.

After the preparation of NiOOH, invented by *Jungner* and *de Michalowski*, in 1901 *Waldemar Jungner* (Sweden) and *Thomas Edison* (USA) invented the first nickel–cadmium (Ni–Cd) and nickel–iron (Ni–Fe) batteries and collaboratively applied for a patent. The whole redox reaction during the current flow in the Ni–Cd battery with nickel as the positive electrode (cathode) and cadmium as the negative (anode) is given by:

$$Cd + 2NiOOH + 2H_2O \leftrightarrows Cd(OH)_2 + 2Ni(OH)_2 \tag{1.5}$$

For many years, Ni/Cd served as the best and only option in the rechargeable battery for several applications. However, because of the toxicity of cadmium, a search for new materials in order increase the energy density has been ongoing.

A significant step towards giant battery technology was made following the use of Li ions in batteries. Among all metals, lithium was found to be lighter, had high electrochemical potential, high theoretical specific capacity, and hence was a good choice as a negative electrode to improve the energy density of a battery. In 1991, the Sony industrial group from Japan developed the first commercialized lithium-ion battery. Before that, *Yazami* explored reversible intercalation of lithium in graphite and *Goodenough* achieved successful synthesis of LCO ($LiCoO_2$), a layered structure material, which was further found to be a potential candidate as a positive electrode material in LiBs. In 1997, *Goodenough* at the university of Texas (USA) invented another efficient material, lithium iron phosphate ($LiFePO_4$), as a positive electrode for LiBs. Since the last two decades, many advances have been made with new kinds of materials and their combinations for further development in LiB technology with high energy density, power density, energy efficiency, and cyclic stability. However, there is still huge room for development to obtain optimum performance of batteries under all operating conditions.

1.3 Fundamentals of ECCs and battery metrics

1.3.1 Components of ECCs

The anode, cathode, and electrolyte are the three fundamental parts of any electrochemical cell. The electrodes are electrical conductors whereas the electrolyte is an ionic conductor; nevertheless, the anode and cathode are never made of the same conducting materials. The electrodes are separated by immersing them in an electrolyte solution. A typical ECC structure is shown in figure 1.1.

Anode: A negative electrode to which anions (negatively charged ions) migrate, i.e. the anode donates electrons to the external circuit as the cell discharges. It is generally a metal or an alloy.

Cathode: A positive electrode to which cations (positively charged ions) migrate. It is usually a layered structure of metallic oxide, sulfides, and or carbon material.

Electrolyte solution: It is a solution containing dissociated salts, which enable ion transfer between the two electrodes.

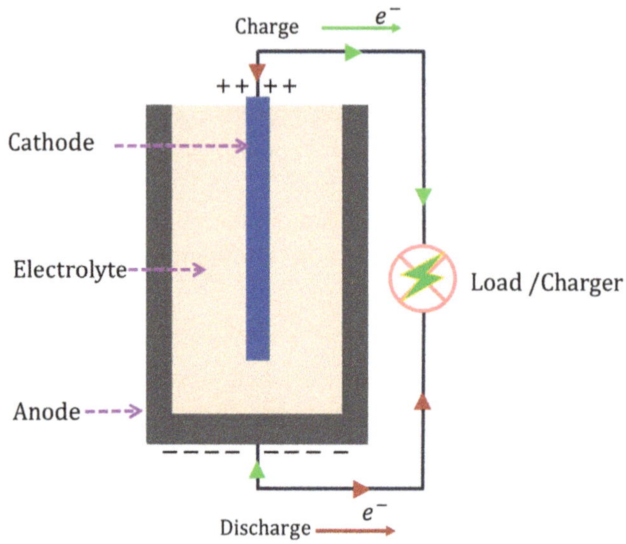

Figure 1.1. Schematic of an electrochemical cell.

1.3.2 What makes an anode (−ve) and a cathode (+ve)

The redox reactions at the interfaces between the electrolyte and electrodes are accompanied by the diffusion of ions in the electrolyte. Ions in the electrolyte react with atoms in the anode and this results in the build-up of electrons causing the anode to be negatively charged. While at the cathode, chemical reactions with electrolytes cause electrons to be used up, which results in the cathode becoming positive. To make the anode electron rich, the external power source is connected between the anode and the cathode, known as charging of the cell. The electrolyte acts as a barrier for the movement of electrons directly from the anode to cathode within the cell. Discharging can be done using an external circuit by connecting electrodes through a load thereby liberating electrons, and allowing a current to flow through the circuit. Current rechargeable batteries are based on the ion insertion phenomena in the electrode material matrix, which allows them to undergo several cycles through charge and discharge operations. Electrochemical redox processes at the electrode surface are also involved in the charging and discharging of batteries. Through these reactions, electric energy is converted into chemical energy and vice versa.

The complete redox process can be divided into two half-reactions, with one half being oxidation (electrons are lost at the anode) and the other reduction (electrons are gained at the cathode). Therefore, it can be stated simply that throughout the reaction, the anode is oxidized and the cathode is reduced. There is a distinct standard potential for each half reaction.

When an electron goes from the cathode to the anode during charging, it transforms electrical energy into chemical energy, increasing the chemical potential energy. The chemical potential energy decreases when the electron passes from the anode to the cathode during discharging, converting chemical energy to electrical energy.

1.3.3 Some battery terminology [4, 5, 7, 8]

Energy storage system: It basically refers to a battery pack system, meaning an electrical or mechanical combination of ECCs with appropriate thermal, electrical and mechanical specifications.

Intercalation: A process of inserting a guest ion in the host matrix. For this the host must have a layered structure. In the case of a Li-ion battery, the guest is the Li ion and the host is the layered electrode material.

De-intercalation: The process of taking out a guest ion from the host matrix.

Capacity: Measure of total energy available with the battery or total charge stored in a battery, measured in ampere-hour (Ah). Ampere-hour is the capacity with the battery. It is basically the current that the battery can provide over a specified time period. So, the larger the current the more power can be released. Thus, according to the definition, a 10 Ah cell is able to supply 10 A for a 1 h period. But, according to the system specification, the rate with which the power is delivered by the battery may vary. (Example: a battery that delivers 10 A for 10 h delivers $10\ A \times 10\ h = 100\ Ah$ of capacity.)

Cell voltage: Cell voltage is represented by open-circuit voltage or working voltage, i.e. closed-circuit voltage.

Open-circuit voltage: This is the voltage between the positive and the negative electrodes when no external current flows (i.e. for no load condition). It is calculated by comparing the chemical potentials of the electrodes. To get it as high as possible, the chemical potential of positive electrode should be more than that of negative electrode. Also, the redox energies of the electrode should lie within the band gap of the electrolyte used.

Working voltage: The closed-circuit voltage between the positive and the negative electrodes connected through the external load.

Discharge rate or C-rate: The rate at which a battery is discharged relative to its maximum capacity. It is described in relation to the time of 1 h discharge.

Consider an example of a battery with capacity: 1000 mAh = 1 Ah.

For such a battery, the C-rate means that the entire battery is fully discharged (or charged) in 1 h. Likewise, a 2 C-rate means that the entire battery is fully discharged in 0.5 h, i.e. 30 min (1 h/2 C = 60 min/2 C = 30 min). A 1/5 C-rate would be equal to a 5 h discharge period (1 h/(1/5) C = 5 h).

A 0.05 or 1/20 C-rate would be equal to a 20 h discharge period. Thus, as the C-rate increases, the discharge time goes down and vice versa. This is a very important factor in order to evaluate the power output ability of the energy storage system. Some high-performance batteries can be charged and discharged above 1 C-rate with moderate stress.

Many EV systems require a continuous current draw that may be from 20 A to 400 A. In order to achieve such a requirement one can either use a single cell or lower Ah cells in a parallel configuration. For example, if the specification is 100 Ah at a 1 C-rate, this means that a battery can have 10 A of capacity over a 1 h period.

The same cell can be rated up to a 5 C-rate for a period of 10 s, which can provide 500 A (100 A × 5) over 10 s.

Nominal voltage: Average voltage during the total discharge process of a battery at the rate of 0.2 C.

Nominal capacity: The total capacity during the discharge process of a battery at the rate of 0.2 C.

Discharge capacity: The total number of electrons transferred during a discharge process. A 3600 coulomb charge corresponds to a 1 Ah discharge capacity.

Depth of discharge (DOD): Measure of how much of the percentage of battery capacity can be used relative to its total capacity for a particular application to avoid over discharge.

It is the percentage of decrease in the maximum capacity of a battery during the discharge process; 80% DOD is referred to as a deep discharge. For greater cycle life and safety of the storage system, DOD is kept at minimal value during the design.

State of charge (SOC): A measure of how much capacity or power is left in a battery. Thus, it is opposite to DOD. DOD measures how much of the battery can be used, while SOC measures how much is left after a specified period of time for a particular application. It is always measured against DOD.

Duration time: The total time required by a battery to discharge until terminal discharge voltage.

Cut-off/terminal voltage: The final voltage between two electrodes during the complete charge or discharge process.

Self-discharge: Due to internal chemical reactions in a cell a decrease in stored capacity occurs even in the open-circuit condition, known as self-discharge. It causes a decrease in the shelf life of the battery.

Shelf/storage life: The best possible valid time that a battery can be stored or preserved without any load.

Cycle: The process of complete discharge and then charge is known as the cycle for a battery.

Cycle life: The number of times that a battery can be recharged or cycled, i.e. charged and discharged.

Over discharge: Occurring when a discharge voltage is below the specified terminal voltage value.

Over charge: Occurring when a charge voltage raises the specified terminal voltage value. It causes the cell to stop working due to emission of gases by decomposition of the electrolyte.

Constant voltage charge: A constant potential maintained during a charging process. When the battery voltage arrives at the specified voltage, this process terminates.

Constant current charge: A constant current maintained during a charging process. When the battery current arrives at the specified capacity, this process terminates.

Power efficiency: The ratio of the energy expended by all the external circuit components compared with the battery energy consumption.

1.3.4 Performance measuring key battery attributes

The datasheet of a battery always comes with some standard terms to represent its performance quality. The most common parameters that are used to validate the quality storage system are:

- Cell voltage;
- Discharge rate/C-rate;
- Specific capacity;
- Capacity retention (stability/cycle life);
- Energy density: gravimetric and volumetric energy density;
- Power density;
- Coulombic efficiency;
- Cost, toxicity, and safety issues.

1.4 Lithium-ion battery

Li-ion batteries (LIBs) are a form of rechargeable battery made up of an electrochemical cell (ECC), in which the lithium ions move from the anode through the electrolyte and towards the cathode during discharge and then in reverse direction during charging [8–10].

It was invented in 1991 by the Sony corporation for portable telephones with lithium–cobalt oxide ($LiCoO_2$) as the positive electrode material and carbon as the negative electrode. The cell produced an electrochemical capacity of about 160 mAh g^{-1} [11]. For the development of lithium-ion batteries in 2019, *John Goodenough, Stanley Whittingham, and Akira Yoshino* received the Nobel Prize in Chemistry.

Due to their high energy density, long cycle life, high open-circuit voltage, and low self-discharge rates, lithium batteries have now been conclusively shown to be the finest secondary batteries available. However, due to numerous complex phenomena at each stage, from material synthesis to device assembly, the creation of new high-energy lithium-ion batteries is a promising job. To sustain the steady advancement of high-energy lithium battery systems, a systematic scientific approach and a development plan for new anodes, cathodes, and non-aqueous electrolytes are required.

1.4.1 Importance of lithium metal in battery technology

Lithium is the third simplest element, with only three electrons, after hydrogen and helium. In comparison to lead and zinc in conventional batteries, lithium has a substantially higher energy density. It offers the highest specific energy per weight and the highest electrochemical potential. Additionally, molecular mechanisms, such as how lithium can mix with carbon to generate lithium carbonate, are well understood.

There are three key benefits of lithium for batteries:

1. First, it is highly reactive because it readily loses its outermost electron and facilitates current flow via batteries.
2. Second, it is much lighter than other metals used in batteries.
3. Third, it is also a highly reactive element, meaning that a lot of energy can be stored in its atomic bonds, which translates into very high energy density for lithium-ion batteries.

1.4.2 Components of a LIB

Like other ECCs, a rechargeable LIB is also made of the same components with one or more cells together. Figure 1.2 illustrates the components of a LIB.

The basic structure comprises:

- Cathode: positive electrode made up of lithium metal oxide as the cathode on aluminum foil;
- Anode: negative electrode made up of carbon as the anode on copper foil;
- Electrolyte: lithium salt in organic solvent;
- Separator: made up of polyethylene or propylene.

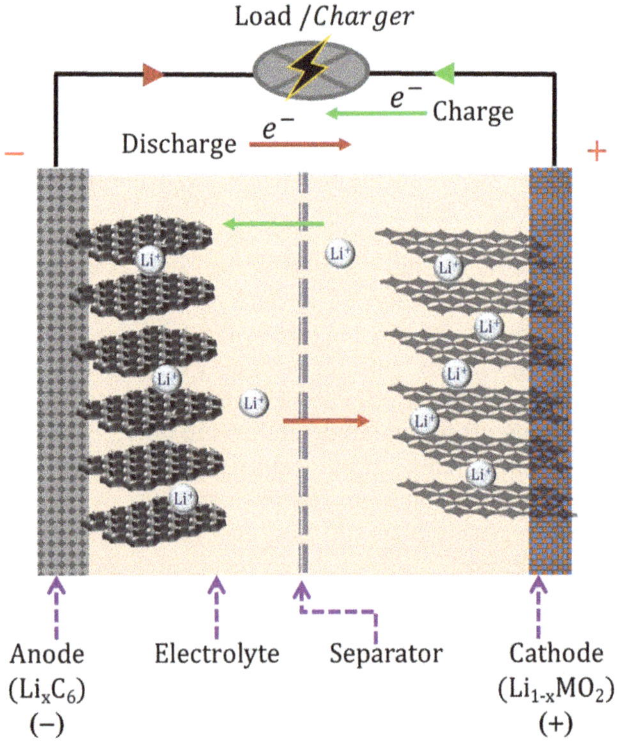

Figure 1.2. Components of a LIB.

During charging, the Li ion moves from the cathode to the anode through the electrolyte and electrons through the external circuit. Thus, the external power is used to store energy chemically. During energy utilization, i.e. discharging, the electrons move from the anode to cathode through the external circuit and at the same time the Li ions move back to the cathode via the electrolyte. Thus, LIBs are rechargeable due to the ease with which lithium ions and electrons can be transferred back into negative electrodes. A separator is used to avoid direct contact of the electrodes and only allows the working ion to freely pass through it [10–14].

The positive electrode, i.e. cathode, is typically made from a chemical compound called layered lithium metal oxide, for example: lithium–cobalt oxide ($LiCoO_2$), and the negative electrode, i.e. anode, is generally made from carbon/graphite compounds [12].

The cathode material that stores lithium ions via electrochemical intercalation must contain suitable lattice sites to store and release ions reversibly, hence material with layered structures may offer stable cyclability and high specific capacity. In addition to this, differential electrochemical potential between the cathode and anode is necessary to obtain a high energy density battery with a given anode. The role of the electrolyte is to act as a medium for the transfer of ions between the two electrodes and to block the electrons.

1.4.3 Battery charging and discharging process

Figure 1.3 shows a schematic representation of the charge and discharge processes in LIBs using $LiCoO_2$ and graphite as typical electrode materials [15].

During charging, the LCO positive electrode gives up some of its lithium ions, which move through the electrolyte towards the negative, carbon/graphite electrode and remain there. Electrons also flow from the positive electrode to the negative electrode through the external circuit. The electrons and ions combine at the

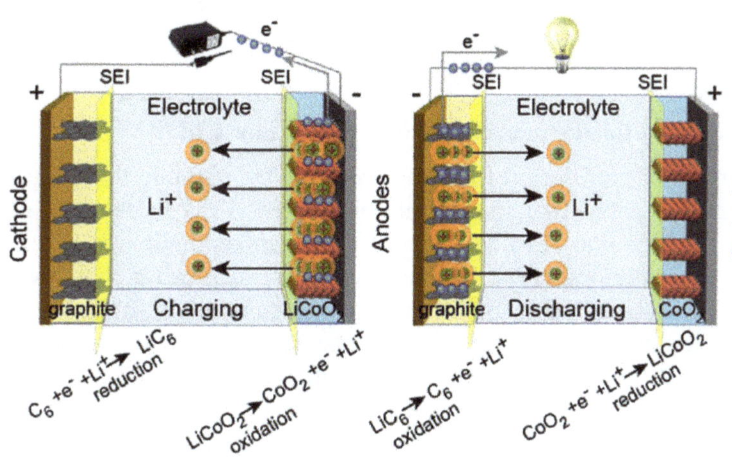

Figure 1.3. Schematic illustration of the charging and discharging processes in LIBs using graphite and $LiCoO_2$ as electrode materials. Reproduced from reference [15] with permission from the Royal Society of Chemistry.

negative electrode and deposit lithium there. Once the moment of most of the ions takes place, decided by the capacity of the electrode, the battery is said to be fully charged and ready to use. When the battery is discharging, the lithium ions move back across the electrolyte to the positive electrode (the $LiCoO_2$) from the carbon/graphite, producing the energy that powers the battery. In both cases, electrons flow in the opposite direction to the ions around the external circuit. Electrons do not flow through the electrolyte: it is effectively an insulating barrier, so far as electrons are concerned.

During the whole cycle of charge and discharge, Li^+ extracts from the cathode and intercalates into the anode and swims back from the anode to the cathode. During charging, the half reaction at the positive electrode represents oxidation and another half reaction at the cathode represents reduction. Overall, during charging, Li^+ flows from the $LiCoO_2$ cathode to the graphite or carbon anode (where it gets intercalated) through the electrolyte, which results in the oxidation of Co^{3+} to Co^{4+}. The opposite occurs during discharge, i.e. Li^+ moves from carbon to $LiCoO_2$. The process is completely reversible.

Battery voltage is the result of the significant difference between Co oxidation and graphite reduction during the moment and Li^+ ion intercalation. Thus, during charging and discharging, lithium ions move back and forth between the electrodes. The reaction mechanism is described by equations (1.6), (1.7) and (1.8) [12, 16].

$$\text{On cathode}(+)\text{: } Li_{1-x}CoO_2 + xLi^+ + xe^- \xLeftrightarrow{\text{Charge/discharge}} LiCoO_2 \tag{1.6}$$

$$\text{On anode}(-)\text{: } xLiC_6 \xLeftrightarrow{\text{Charge/discharge}} xLi^+ + xe^- + xC_6 \tag{1.7}$$

$$\text{Overall: } LiCoO_2 + xC_6 \xLeftrightarrow{\text{Charge/discharge}} Li_{1-x}CoO_2 + xLiC_6 \tag{1.8}$$

In conclusion, the electrical nature of an electrode is determined by the diffusion of ions and redox processes at the electrolyte/electrode interfaces.

1.4.4 Driving force for the moment of lithium ions in a LIB

For electrons to be transferred from the anode to the cathode there must be some sort of energy potential that makes this phenomenon favorable. The gradient of electrochemical potential of Li between the anode and cathode in a LIB is the main driving force for working LIBs. Li ions shuttle like a 'rocking chair' between two electrodes. The concentration of lithium ions remains constant in the electrolyte regardless of the degree of charge or discharge, it varies in the cathode and anode with the charge and discharge states. The potential energy that drives the redox reactions involved in the electrochemical cells is the potential for the anode to become oxidized and the potential for the cathode to be reduced. During discharge, Li^+ ions de-intercalate from the anode and intercalate into the cathode. To maintain electrical neutrality, electrons leave the anode simultaneously and flow into the cathode through the external circuit, which provides electrical energy to the load.

The general effect of discharge is to convert chemical energy into electrical energy by means of the redox reaction of electrodes driven by the chemical potential difference between the cathodes and anodes. The opposite is true during charging.

1.4.5 Fundamental principle of LIB electrochemistry [17–20]

The electrochemical potentials of the electrode materials are a significant contributing component and can aid in the design and development of appropriate intercalation materials to produce batteries with a high energy density. In terms of several electrochemical characteristics, such as battery voltage, specific capacity, specific energy, specific power, energy density, coulombic efficiency, etc, the performance of a LiB can be assessed. It is possible to estimate the theoretical values of each of these parameters, which correspond to the material's equilibrium performance. However, due to a variety of complicated concerns from a scientific and technological point of view, the values in practice may deviate from the estimated theoretical values.

Based on a deep understanding of battery electrochemistry and related studies, the performance reducing issues in batteries can be broadly divided into the following categories:

- Charge transfer and mass transport process (both in bulk and at the interface);
- The structural change and phase change of the electrode during the Li ion de-intercalation;
- Relationship between the electrical and chemical effects at time scale.

Battery voltage:
The battery voltage is the driving force (thermodynamically, the electrochemical potential difference) pushing alkali ions and electrons from one electrode to the other. Aydinol *et al* [18] proposed the mechanism of battery voltage calculation, considering the system as a thermodynamic system. According to the Nernst equation and the second law of thermodynamics, the potential is proportional to Gibbs free energy.

$$\Delta G = nFV \tag{1.9}$$

Here, ΔG is the change in Gibbs free energy, for the reaction during the charge and discharge state, n is the total number of electrons transferred (depending on the 'z', valance of the working ion) and F is Faraday's constant, 96 485.3 C mol^{-1}.

The Gibbs free energy is related to the total internal energy, temperature, and entropy. Thus,

$$\Delta G = \Delta H - T\Delta S \tag{1.10}$$

where H is enthalpy, i.e. the total amount of energy of a thermodynamic system, T is temperature, and S is entropy.

Comparing both equations,

$$V = \frac{\Delta H - T\Delta S}{nF} \tag{1.11}$$

This is an ideal open-circuit voltage of a battery, which is temperature dependent.

$$\text{But } \Delta H = \Delta U_{\text{int}} - P\Delta V \tag{1.12}$$

where ΔU_{int} is the change in internal energy, P is pressure, and ΔV is change in volume.

Therefore,

$$\Delta G = \Delta U_{\text{int}} - P\Delta V - T\Delta S \tag{1.13}$$

$$\Delta G = \Delta U_{\text{int}} - (P\Delta V + T\Delta S) \tag{1.14}$$

The terms $P\Delta V$ and $T\Delta S$ correspond to the structural volume and configurational entropy change due to ion intercalation and de-intercalation.

At $T = 0$, other thermodynamic parameters have negligible contribution to G as compared to U_{int}, and hence can be neglected.

Therefore,

$$\Delta G = \Delta U_{\text{int}} \tag{1.15}$$

Thus, based on the concentration of the intercalated ion (x), the voltage of a battery can be calculated in terms of a change in Gibbs free energy due to the Li^+ intercalation process, and is represented as

$$V(x) = \frac{\Delta G(x)}{\Delta x} = \frac{\Delta U_{\text{int}}(x)}{\Delta x} \tag{1.16}$$

where Δx is the change or transfer of Li^+ concentration during the intercalation process.

Thus, the equilibrium potential of a battery is given by:

$$V(x) = -\frac{\Delta G(x)}{nF} \tag{1.17}$$

But, in the case of a battery, G is derived from the combined energy of the cathode (G_C) and anode (G_A) at one state of charge in relation to some initial concentration, x_0. The total number of electrons transferred (n) depends on the valance of the working ion (z) and F is Faraday's constant.

Thus,

$$\Delta G(x) = [G_C(x) - G_C(x_0)] - [G_A(x) - G_A(x_0)] \tag{1.18}$$

and $n = z(x - x_0)$.

Therefore, the voltage can be

$$V(x) = -\frac{\{[G_C(x) - G_C(x_0)] - [G_A(x) - G_A(x_0)]\}}{z(x - x_0)F} \tag{1.19}$$

Using a definition of chemical potential ($\mu = \partial G/\partial x$)

$$\text{Thus, } \mu_A(x) = \frac{[G_A(x) - G_A(x_0)]}{(x - x_0)} \text{ and } \mu_C(x) = \frac{[G_C(x) - G_C(x_0)]}{(x - x_0)} \tag{1.20}$$

$$V(x) = -\frac{[\mu_C(x) - \mu_A(x)]}{zF} \tag{1.21}$$

Thus, the equilibrium cell potential is the theoretical voltage of a battery that depends on the difference between the chemical potential of lithium in the anode and cathode material. This equilibrium potential is equal to the open-circuit voltage (V_{oc}) when no current is passing through the external circuit, i.e. under no load condition.

It is possible to determine the theoretical voltage in lithium intercalation compounds (for instance, layered lithium transition metal oxides $LiMO_2$) generated as a result of the insertion or extraction of Li ions as follows:

$$\Delta G = G_{Li_{x_1}MO_2} - G_{Li_{x_2}MO_2} - (x_2 - x_1)G_{Li} \tag{1.22}$$

$$V = -\Delta G/(x_2 - x_1) \tag{1.23}$$

The number of electrons or ions transferred $n = (x_2 - x_1)$,

where $x_2 - x_1$ is the concentration of the intercalated Li ion.

However, the operating voltage of the battery is also constrained by the electrochemical window of the electrolyte, which is the energy gap between the highest occupied molecular orbital (HOMO) and the lowest unoccupied molecular orbital (LUMO).

The following variables are used to compare and describe the performance of battery: specific capacity, specific energy, specific power, energy density, cycle life, and coulombic efficiency.

Specific capacity: The specific capacity of a battery is the number of electrons delivered per unit mass of electrode material. The maximum specific capacity of intercalation electrodes is determined by the number of electrons intercalated or de-intercalated during cycling and the molecular weight of the insertion material. It is given by

$$C_{max} = \frac{nF}{3600\,M} \tag{1.24}$$

$$C_{max} = \frac{nF}{D} \tag{1.25}$$

where D stands for the constituent element's atomic weight in the cathode materials.

The number of available sites for active M^{n+} ions that are intercalated into the host matrix limits the capacity of a battery. In practice, the specific capacity is reported with respect to the mass or volume of the active material, accordingly known as gravimetric specific capacity (mAh g^{-1}) or volumetric specific capacity (mAh cm^{-3}), respectively. The charge storage capability measured per unit area is referred to as areal capacity and represented in mAh cm^{-2}.

The volumetric specific capacity is important when designing a battery with certain specifications useful for high energy applications like electric vehicles and grid storage.

Energy density: The total energy stored in a battery can be calculated by integrating voltage with respect to its capacity. It is referred to as energy density and is related to working voltage and reversible capacity.

$$E = \int V(C)\mathrm{d}C = \int V(t)I\mathrm{d}t \tag{1.26}$$

Working voltage decided by the potential of redox processes in a cell and the reversible capacity depends on the amount of lithium ion intercalated. Thus, to ensure the highest energy density the redox pair should have higher potential and the electrode should have a layered structure with a variety of material compositions. Energy density can be referred to as gravimetric energy density if measured in watt hour per kilogram (Wh kg^{-1}) and as volumetric energy density if measured in watt hour per liter (Wh L^{-1}).

Power density: It is the maximum power that can be delivered by the battery with respect to its mass. It is also determined as the maximum rate of discharge energy per mass or volume of a battery. Thus, the energy density defines how much energy is supplied by the battery to do the work, while power density defines how fast the work can be done with the available energy.

The battery power can be written as:

$$P = I\ V - I^2\ R_{\mathrm{int}} \tag{1.27}$$

where I is the current drawn from a battery, V is the battery voltage, and R_{int} is the internal resistance of the battery. Thus, to get maximum power the internal resistance must be kept to a minimum. The value of R_{int} depends on the ionic conductivity of alkali ions, electrical conductivity of the electrode material and reaction kinetics during intercalation and de-intercalation. It is usually categorized as gravimetric power density if represented as watt per kilogram (W kg^{-1}) and volumetric power density if expressed as watt per liter (W L^{-1}). It can be determined by cycling the battery at different charging/discharging rates.

To get an optimized balance between energy and power density, chemical/thermal stability and, more specifically, surface reactivity of electrode materials play a vital role. Better performance cannot always be obtained by searching for new materials, but one can also enlist known materials by optimizing their properties such as conductivity, surface morphology, electrochemical, chemical, and structural stability.

Coulombic efficiency: The ratio of energy withdrawal from a battery during discharge to the energy used during charging of a battery. In other words, it is the ratio of charge extracted to charge inserted to the battery over one cycle. It is also known as charge/discharge efficiency. It has a value near to about 100%.

Voltage/voltaic efficiency: It is the ratio of average discharge voltage to the average charge voltage of a battery. An electrode material with high active surface area and high electrical conductivity gives more voltage efficiency.

Energy efficiency: It is the ratio of energy densities during the discharge and charge process. It is greatly influenced by the charge and discharge rate.

Electric efficiency: It is the product of voltage efficiency and coulombic efficiency.

Charge factor: The reciprocal of coulombic efficiency is known as the charge factor.

Cycle life: The number of times a battery can be charged and discharged with a given C-rate. According to industry standards, the cycle life of a battery is almost over when the specific capacity reaches 80% of its initial value after certain number of charge/discharge cycles. As stated earlier, the depth of discharge (DOD) measures how much of the battery capacity can be used for application to avoid over discharge. Therefore, as the cycle life significantly depends on DOD, it is necessary to report it with respect to depth of discharge. The cycle life of a battery also depends on several other factors such as operating temperature, rate of charge or discharge, charge/discharge cut-off voltage, and storage condition.

The cycle life, energy density, power density, and rate capability of a battery mainly depend on the electric and ionic conductivities of the electrode materials. Thus, to obtain excellent performance and to maintain the low internal resistance of a battery, higher electric conductivity electrode materials are useful. Also, low molar mass electrode materials are required to ensure higher capacity and rate capability.

1.5 The pros and cons of LIBs [13, 19, 21–23]

Compared to other secondary batteries, LIBs have remained in existence for a long time at the top locus in the majority applications due to their superior energy storage performance. At present, almost all sectors where low or high power is needed such as consumer electronics, smart grid energy storage, electric vehicles, and even for aerospace, regard LIBs as one of the best candidates. Everything comes with certain pros, cons, and limitations, and LIBs are no exception. Understanding their advantages, disadvantages, and limitations will significantly enable technological advancement.

Advantages

The advantages of LIBs include:

LIB electrochemistry is more efficient than other secondary batteries

There are numerous electrode and electrolyte combination options available with LIBs. Research and development in new cathode, anode, electrolyte, and electrochemical activities in LIBs has advanced significantly during the last 20 years. Since invention of LIBs, theoretical and experimental research has boosted power density, energy density, and life expectancy. For instance, the LIB has roughly twice the energy density of a typical Ni–Cd battery.

Low self-discharge rate

Even in the absence of load, a battery experiences chemical reactions that induce self-discharge, which is a certain charge loss. The LIB exhibits only a small amount of self-discharge, which is only about 5% in the first four hours after charging and thereafter only 1–2% per month. The self-discharge of a LIB battery is half that of a Ni–Cd battery.

Easy maintenance
The LIB does not need regular active maintenance like lead–acid batteries, and it has a portable design and one-time purchase warranty. Its cycle life is ten times greater than that of lead–acid batteries, and over 2000 cycles, it performs at about 80% of rated capacity.

Environmentally friendly and sustainable
In comparison to lead–acid and Ni–Cd batteries, LIBs contain comparatively few hazardous elements. Additionally, LIBs can be charged quickly and efficiently at any place with basic chargers, whereas lead–acid batteries need to be charged over an extended period of time.

In comparison to a lead–acid battery, the LIB offers more energy in only half the mass. As a result, it uses less material, is smaller, and is better suited for easy installation. For instance, a typical LIB has a storage capacity of 150 watt-hours per kg, compared to perhaps 100 watt-hours for nickel–metal hydride batteries. However, a lead–acid battery can store only 25 watt-hours per kg. A lead–acid battery must therefore weigh 6 kg in order to store the same amount of energy as a 1 kg LIB.

No memory effect
If a battery is partially discharged before being recharged, then it will deliver the amount of energy which is used during partial discharge, this is known as the 'memory effect', or 'lazy battery effect'. Lithium-ion batteries don't suffer from memory effect, which means that there is no need to completely discharge before recharging.

High cell voltage
A single cell of a LIB provides a working voltage of about 3.6 V, which is almost two to three times higher than that of a Ni–Cd, NiMH, and lead–acid battery cell.

Good load characteristics
The LIB provides steady voltage under any load condition. It has good working performance until its reasonable discharge, i.e. successfully retains constant voltage per cell.

High energy and power density
Lithium is a highly reactive element, meaning that a lot of energy can be stored in its atomic bonds, which translates into high energy density for lithium-ion batteries. Hence, it can be used in adequate sizes for applications from portable electronic devices, smartphones, to electric vehicles. The use of electrode materials with an effective electrochemical surface area provides reasonable energy and power density. While for applications like electric vehicles, there is an ongoing requirement for batteries with higher power density, and therefore more efforts in this regard are still in progress.

Numerous choices for electrode materials
As a result of global research and development efforts, several classes of nano-structured materials, including lithium metal oxides, chalcogenides, silicates and

phosphates, with different chemical compositions, crystal structures, and higher surface/volume ratios have been revealed to provide additional sites for Li storage. This demonstrates an improvement in the electrochemical performance of LIBs.

Long cycle life
Numerous factors can affect a battery's cycle life. Compared to lead–acid batteries, under standard conditions, with minimal value of DOD, a LIB has a greater cycle life of about 1000–1500 charge/discharge cycles. Also, other secondary batteries have the problem of corrosion due to high DOD.

High round trip efficiency
Round trip efficiency is a measure of the energy retention of a battery after its fully charged state. Due to fewer losses in storage or self-discharge, the LIB has a higher round trip efficiency of about 95% compared to other batteries.

Disadvantages
Despite its benefits, the LIB, like other batteries, has several shortcomings and challenges, including:

- Due to the liquid polymerized electrolytes, further safety procedures are needed;
- Protective covers are usually required to prevent short circuits;
- To avoid short circuits one cannot carry the LIB separately, hence it always requires a protective cover;
- Due to its fragility, every LIB requires a protective circuit;
- Advanced battery management systems are necessary for high power applications, which raises the overall cost;
- Due to its high sensitivity to temperature and potential for explosion upon overheating, the LIB cannot be used in transportable devices;
- Fast charging of most batteries is limited by the physical condition of their surroundings;
- The aging effect is significant in the LIB, it also shows declined performance even if not in use for a long time.

1.6 Overview of the LIB assembly process [24–26]

Like other batteries, LIBs go through a few stages in the manufacturing process that significantly affect their overall performance, durability, cost, and power consumption. The components that structure the basic assembly of a LIB cell are depicted in figure 1.4.

The preparation of the electrode, as well as the processes of assembly, creation, and testing, are all steps in the fabrication of a LIB cell. Figure 1.5 shows a schematic representation of the fabrication process for a LIB cell [24].

Slurry preparation
The active material is mixed with a conductive additive, binder, and solvent in appropriate mass ratios to form a slurry. N-methyl pyrrolidone (NMP) and polyvinylidene fluoride (PVDF) are normally used as the solvent and binder, respectively. Formation of uniform slurry is very important to get appropriate

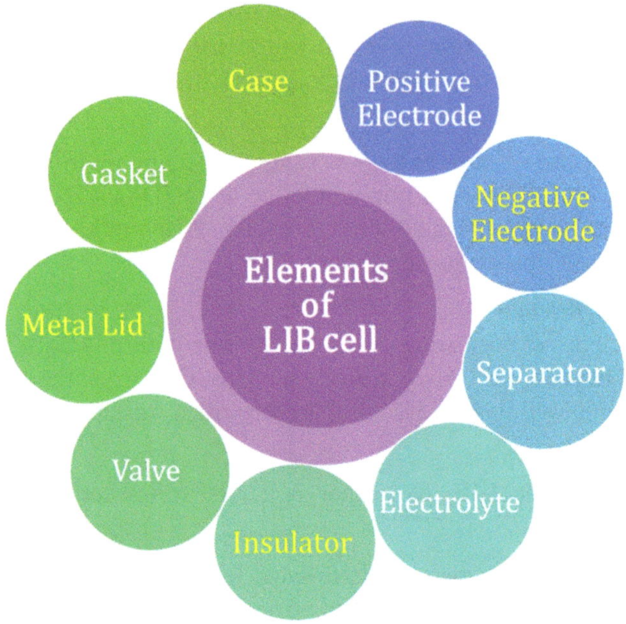

Figure 1.4. The building blocks for a fundamental LIB cell.

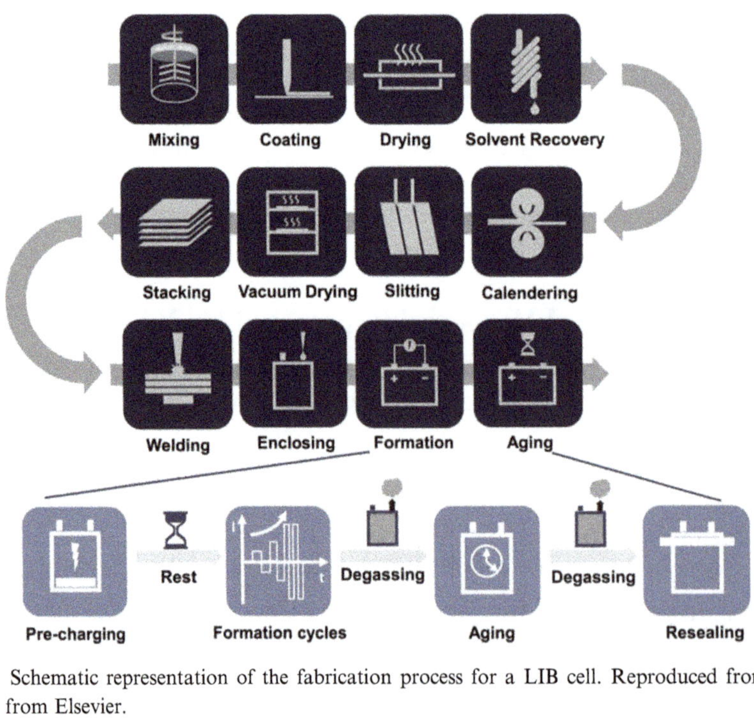

Figure 1.5. Schematic representation of the fabrication process for a LIB cell. Reproduced from [24] with permission from Elsevier.

electrode material distribution on a current collector foil. In LIBs the current collectors are aluminum foil for the cathode and copper foil for the anode.

Electrode coating and drying
After mixing, the uniform slurry is pumped to a slot die and coated on surfaces of the current collector. The conventional roll-to-roll slot die coater system is used for uniform coating of the electrode. The electrode is then subject to drying in an oven to evaporate the solvent. Drying is a very important stage because, during this, the cross-linked continuous network is formed by the conducting additive and binder, which increases the effective electro active sites and conductivity of the electrode.

Calendaring
The coated electrode is then compressed to get the desired thickness. This process is known as pressing or calendaring. This step is very important to increase the bonding strength between the electrode and the current collector. It also helps to optimize the electrode's physical properties such as mass loading, porosity, conductivity, etc.

Most of the research efforts show a preference for two-step roll pressing. Firstly, the electrode undergoes soft roll pressing to ensure uniform distribution of the microstructure and pores. Afterward, it is subjected to forceful roll pressing to obtain the desired density and thickness.

Electrode slitting
According to the size of the battery case, the electrode foils and separator are stamped and slitted or cut using slitting machines. To fit the electrode in the various battery designs accurately, the slitting machine must be very precisely manufactured and calibrated. Laser cutting will be the best option in the near future.

Vacuum drying
In order to remove excess moisture the slitted electrodes are placed in a vacuum oven for drying. Nowadays, the argon purging method for drying is found to be a better option to obtain good electrochemical performance of the battery.

Welding
Depending on the type of battery casing, the electrodes and separator are stacked together or spirally wound to form an internal structure. To connect the electrode structure to the terminals, tabs of the same current collector material are welded on the cathode (Al tab) and anode (Cu tab) electrode. The most common welding methods are spot welding, ultrasonic welding, bolt welding, and laser welding.

Packaging
The electrode structure is then transferred to the designed case or can. The can is filled with electrolyte before sealing or has a facility for injecting the electrolyte into the case. The sealing can be done by laser welding or a heating process, depending on the case material, leaving an opening.

Formation and aging process
The process of warranting or confirming operational stability of a battery is known as cell formation. It is performed to activate the working material, hence it is also known as the battery electrochemistry activation process. Monitoring the effect of

SEI layer formation on aging and battery performance is the main focus to get an indication of battery capacity degradation and safety issues. Stability of the SEI layer can be ensured by multiple charge and discharge cycles at low rate.

1.7 Classification of LIBs by configuration [27, 28]

Based on their shape and the electrolyte they use, lithium-ion batteries can be divided into two groups.

There are three types of LIB depending on the electrolyte used:

- Conventional LIB models: an organic electrolyte;
- Polymer LIBs: a gel polymer electrolyte;
- Solid LIBs: a solid electrolyte.

Additionally, there are four shape-based variants:

- Coin cell batteries;
- Cylindrical type LIBs;
- Pouch type LIBs;
- Prismatic type LIBs.

The construction of the coin, cylinder, prismatic, and pouch shapes is shown schematically in figure 1.6 [27].

Coin cell battery

This is sometimes referred to as a button cell. It is a circular stainless steel disc with lithium foil as an anode. It has a diameter greater than its overall height. These batteries are used to provide power to portable devices like watches, toys, LED

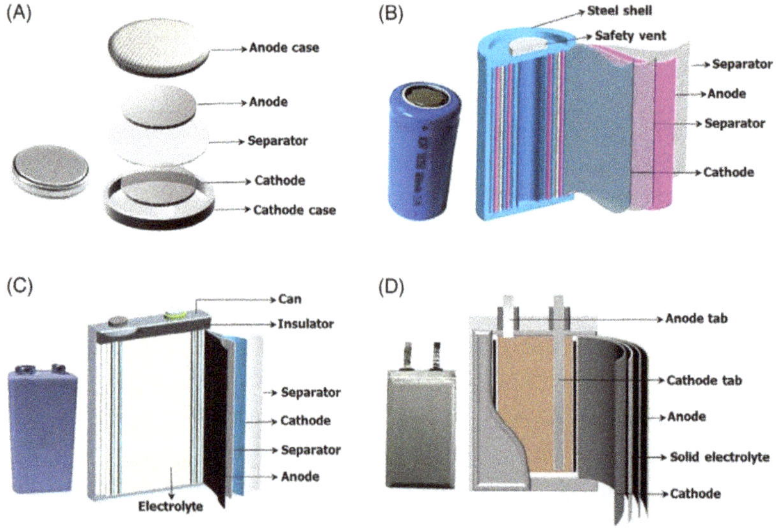

Figure 1.6. Schematic illustration of (a) coin, (b) cylindrical, (c) prismatic, and (d) pouch cell. Adopted from [27], copyright 2019 The Authors. InfoMat published by John Wiley & Sons Australia, Ltd on behalf of UESTC.

lights, health care equipment (thermometers, heart rate monitors, glucose monitors), calculators, remote controls, car keys, etc. The names of coil cell batteries are mostly standardized by the IEC (International Electrotechnical Commission). For example, CR2025 is the most common battery design at laboratory scale, which means it is a single coin cell with a diameter 20 mm and height 2.5 mm. The letter 'C' stands for battery chemistry, here it is Li, 'R' stands for round battery shape, and the number indicates the approximate dimensions of the cell. Several other capital English letters are used to define different shapes of a battery, for example: 'F' stands for 'flat', 'S' stands for 'Square', 'P' stands for 'Not Round', and 'R' stands for 'Round'. With the anode, cathode, and electrolyte the other parts of a coin cell are: case, gasket, cap, plate, and spring. Some examples of coin cell batteries are: CR2477, CR1616, CR1620, CR1632, CR2016, CR2025, CR2032, etc. Figure 1.7 depicts a schematic of the assembly of a coin cell, with all of the components in the sequence and how they are fitted into the cell.

Cylindrical type LIBs
In this type, the electrodes and separator are stacked together and tightly wound spirally in a cylindrical case. It has a solid cylindrical body with flat terminals. It offers good cyclic stability but low packaging density. Cylindrical cells use five

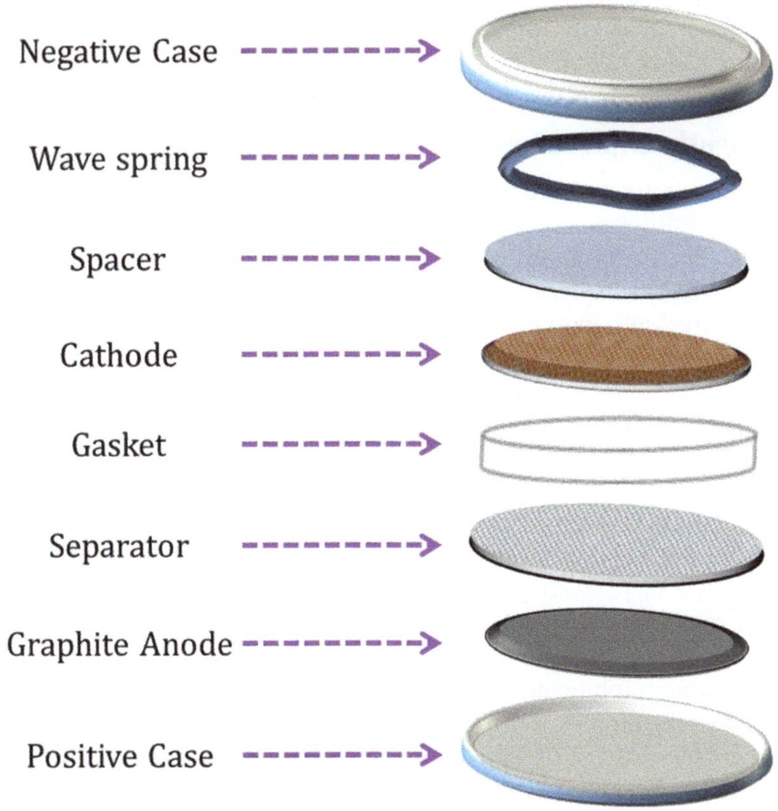

Figure 1.7. Schematic of a coin cell assembly with all of the components.

numbers for marking, for example, 18650 is the most widely used and optimized cylindrical cell, with dimensions 18 mm in length and 65.0 mm in diameter of the cylindrical case. Remember here that the last three digits are for diameter and it is in tenths of millimeters. It is occasionally accompanied with the acronym LIR (lithium ion rechargeable). Several other types of cylindrical batteries are also available such as 14500, 14650, 18500, 18650, 21700, 26650, 32650, etc, and they are commonly used in security and communication systems, medical equipment, laptops, e-bikes, and power tools.

Pouch type LIBs:
A pouch cell has a sealed soft, flexible and flat rectangular pouch type case. The electrodes and separator are stacked instead of wounding. It has light weight, is cost effective, but has a low cycle life.

Prismatic type LIBs:
This type of battery has a metallic or semi-hard plastic case in cubic or rectangular form, which consists of large sheets of electrodes and separator stacked together. It is available with a vent system. Generally, aluminum or steel is used as a metallic case for good stability. These LIBs are attractive in volume-sensitive applications, where the swelling matters significantly. Prismatic cells use six digits for marking, for example, the most common 504050 prismatic cell is 5.0 mm in thickness, 40 mm in width and 50 mm in length. Remember here that the first two digits are for thickness and it is in tenths of millimeters.

1.8 Frequently explored materials in LIB components [29–32]

To date, several materials have been discovered and designed for components of LIBs and continuous efforts in order to improve electrochemical and cyclic performance are still ongoing. The performance of lithium-ion batteries significantly depends on the nature of the electrode material used. Typically, both the cathode and anode in a LIB have layered structures and allow Li^+ to be intercalated or de-intercalated. The most common materials for various components of LIBs are given below:

Cathode

- $LiCoO_2$, Lithium cobalt oxide (LCO);
- $LiMn_2O_4$, Spinel manganese oxide;
- $LiFePO_4$, Lithium iron phosphate (LFP);
- $LiMn_2O_4$, Lithium manganese oxide (LMO);
- $LiNiMnCoO_2$, Lithium nickel/manganese/cobalt oxide (NMC);
- $LiNi_{0.8}Co_{0.15}Al_{0.05}O_2$, Nickel cobalt aluminum oxide (NCA);
- Layered dichalcogenides.

Anode

- Carbon, graphitic and non-graphitic carbon;
- Silicon-based alloys;
- Tin based alloys: Cu–Sn (Cu_6Sn_5), Ni–Sn (Ni_3Sn_2), Co–Sn (Co_3Sn_2), and Sn–Ag alloy;
- Transition metal oxides: Titanium-based anodes ($Li_4Ti_5O_{12}$ and TiO_2).

Electrolyte

- Salt of lithium hexafluorophosphate ($LiPF_6$), or $LiBF_4$, lithium perchlorate ($LiClO_4$), lithium hexafluoroarsenate ($LiAsF_6$) in a mixture of organic solvents (propylene carbonate (PC), ethylene carbonate (EC). Dimethyl carbonate (DMC), diethyl carbonate (DEC), and ethyl methyl carbonate (EMC));
- In the case of a lithium polymer battery, gel electrolyte is used, which involves a polymer such as: polyethylene oxide (PEO), polyacrylonitrile (PAN), polyvinylidene fluoride (PVDF), poly methyl methacrylate (PMMA).

Separators

- Microporous polymeric membrane (polypropylene) or non-woven fabric mats or ceramic based material;
- poly(vinylidene fluoridehexafluoropropylene) (PVDF-HFP);
- poly(vinylidene fluoride-cochlorotrifluoroethylene) (PVDF-CTFE);
- poly(vinylidene fluoride tetrafluoroethylene) (PVDF-TFE).

Current collector

- For the cathode: metallic aluminum foil, mesh, foam etched, or coated;
- For the anode: metallic copper foil, mesh, foam etched, or coated.

1.9 Future strategies for creating next-generation LIBs

The next-generation LIBs can be realized by designing and developing innovative materials for each component to obtain higher capacity, high energy and power density, cyclic stability, longer lifetime, enabling them to be environmentally friendly and cost effective. There is a need to develop novel electrode materials with good electrochemical stability and higher capacities. The layered Li-rich cathodes and silicon based anodes, electrochemically stable and more ion conductive electrolytes, and titanium carbide (MXene) based lighter and cheaper current collectors with flexible structures, represent better hope for greater improvement in the electrochemical performance of LIBs. Recycling of LIBs has been regarded as a promising area to reduce e-waste, protect the environment, and also to provide a secondary source for component materials. However, knowledge about the recyclability processes and subsequent characterization of materials to ensure their quality for further use in batteries needs to be investigated in depth.

References

[1] Pollet B G, Staffell I and Shang J L 2012 Current status of hybrid, battery and fuel cell electric vehicles: from electrochemistry to market prospects *Electrochim. Acta* **84** 235–49

[2] Goodenough J B 2014 Electrochemical energy storage in a sustainable modern society *Energy Environ. Sci.* **7** 14–8

[3] Abruña H D, Kiya Y and Henderson J C 2008 Batteries and electrochemical capacitors *Phys. Today* **61** 43–7

[4] Scrosati B 2011 History of lithium batteries *J. Solid State Electrochem.* **15** 1623–30

[5] Yoshino A 2012 The birth of the lithium-ion battery *Angew. Chem. Int. Ed.* **51** 5798–800

[6] Heth C L 2019 Energy on demand: a brief history of the development of the battery *Substantia* **3** 77–86
[7] Clark S *et al* 2022 Toward a unified description of battery data *Adv. Energy Mater.* **12** 2102702
[8] Zubi G, Dufo-López R, Carvalho M and Pasaoglu G 2018 The lithium-ion battery: state of the art and future perspectives *Renew. Sustain. Energy Rev.* **89** 292–308
[9] Etacheri V, Marom R, Elazari R, Salitra G and Aurbach D 2011 Challenges in the development of advanced Li-ion batteries: a review *Energy Environ. Sci.* **4** 3243
[10] Deng D 2015 Li-ion batteries: basics, progress, and challenges *Energy Sci. Eng.* **3** 385–418
[11] Ozawa K 1994 Lithium-ion rechargeable batteries with $LiCoO_2$ and carbon electrodes: the $LiCoO_2$/C system *Solid State Ionics* **69** 212–21
[12] Ohzuku T and Brodd R J 2007 An overview of positive-electrode materials for advanced lithium-ion batteries *J. Power Sources* **174** 449–56
[13] Blomgren G E 2016 The development and future of lithium ion batteries *J. Electrochem. Soc.* **164** A5019
[14] Ue M, Sakaushi K and Uosaki K 2020 Basic knowledge in battery research bridging the gap between academia and industry *Mater. Horizons* **7** 1937–54
[15] Gulzar U, Goriparti S, Miele E, Li T, Maidecchi G, Toma A, De Angelis F, Capiglia C and Zaccaria R P 2016 Next-generation textiles: from embedded supercapacitors to lithium ion batteries *J. Mater. Chem.* A **4** 16771–800
[16] Borah R, Hughson F R, Johnston J and Nann T 2020 On battery materials and methods *Mater. Today Adv.* **6** 100046
[17] Liu J, Wang J, Xu C, Jiang H, Li C, Zhang L, Lin J and Shen Z X 2018 Advanced energy storage devices: basic principles, analytical methods, and rational materials design *Adv. Sci.* **5** 1700322
[18] Aydinol M K, Kohan A F and Ceder G 1997 *Ab initio* calculation of the intercalation voltage of lithium-transition-metal oxide electrodes for rechargeable batteries *J. Power Sources* **68** 664–8
[19] Kasnatscheew J, Wagner R, Winter M and Cekic-Laskovic I 2018 Interfaces and materials in lithium ion batteries: challenges for theoretical electrochemistry *Modeling Electrochemical Energy Storage at the Atomic Scale* (Cham: Springer) 23–51
[20] Gao J, Shi S Q and Li H 2015 Brief overview of electrochemical potential in lithium ion batteries *Chin. Phys.* B **25** 018210
[21] Sun Y, Liu N and Cui Y 2016 Promises and challenges of nanomaterials for lithium-based rechargeable batteries *Nat. Energy* **1** 16071
[22] Schipper F, Erickson E M, Erk C, Shin J Y, Chesneau F F and Aurbach D 2016 Recent advances and remaining challenges for lithium ion battery cathodes *J. Electrochem. Soc.* **164** A6220
[23] Masias A, Marcicki J and Paxton W A 2021 Opportunities and challenges of lithium ion batteries in automotive applications *ACS Energy Lett.* **6** 621–30
[24] Liu Y, Zhang R, Wang J and Wang Y 2021 Current and future lithium-ion battery manufacturing *iScience* **24** 102332
[25] Wood D L III, Li J and An S J 2019 Formation challenges of lithium-ion battery manufacturing *Joule* **3** 2884–8
[26] Pettinger K H, Kampker A, Hohenthanner C R, Deutskens C, Heimes H and Hemdt A V 2018 Lithium-ion cell and battery production processes *Lithium-Ion Batteries: Basics and Applications* (Berlin: Springer) 211–26

[27] Liang Y *et al* 2019 A review of rechargeable batteries for portable electronic devices *InfoMat* **1** 6–32

[28] Chen T, Jin Y, Lv H, Yang A, Liu M, Chen B, Xie Y and Chen Q 2020 Applications of lithium-ion batteries in grid-scale energy storage systems *Trans. Tianjin Univ.* **26** 208–17

[29] Chen J 2013 Recent progress in advanced materials for lithium ion batteries *Materials* **6** 156–83

[30] Mishra A, Mehta A, Basu S, Malode S J, Shetti N P, Shukla S S, Nadagouda M N and Aminabhavi T M 2018 Electrode materials for lithium-ion batteries *Mater. Sci. Energy Technol.* **1** 182–7

[31] Kalhoff J, Eshetu G G, Bresser D and Passerini S 2015 Safer electrolytes for lithium-ion batteries: state of the art and perspectives *ChemSusChem* **8** 2154–75

[32] Zhu P, Gastol D, Marshall J, Sommerville R, Goodship V and Kendrick E 2021 A review of current collectors for lithium-ion batteries *J. Power Sources* **485** 229321

Chapter 2

Positive electrode materials for Li-ion batteries

The demand for batteries has increased considerably as a result of the necessity of energy storage technologies for sustainable development. Due to the rising demand that will place significant pressure on battery resources, extensive research is being done on alternatives for prospective low-cost battery technologies. Lithium-ion batteries (LIBs) have gained popularity as a promising energy storage technology that may be used for everything from portable electronics to electric cars. More research is being done in order to choose the most promising options more accurately for the anode, cathode, and electrolyte of the lithium-ion cell. To improve the electrochemical performance and cyclic stability of LIBs, several cathode materials have been investigated. This chapter provides a detailed overview of the numerous cathode materials that have been extensively researched along with their advantages and disadvantages. It also highlights the electrochemical reaction mechanisms involved in Li-ion battery operation. In order to gain additional improvement, modifications to adopted high-capacity cathode materials are being made; these modifications are described for each class of material.

2.1 Introduction

Since their inception, lithium-ion batteries (LIBs) have been widely used as a Li ion based energy storage technology due to their superior electrochemical properties, high capacity, comparative superior energy and power density, superior cycle life, almost zero memory effect, and extensive potential for the development of versatile component materials. In the present era, they have become an essential power source, for portable items to space technology. It is almost at the borderline to replace the pioneering secondary batteries, which have stood as the preferable option for energy storage over the past decades. The revolutionary attempts in research and technology of LIBs have made this possible. Considering the global energy need, and environmental and socio-economic issues, the replacement of conventional energy sources by clean energy technology for energy production and

doi:10.1088/978-0-7503-4881-2ch2

storage needs to be explored. The global market of energy storage devices is \$7.81 billion in 2021 and is expected to grow up to \$26.81 billion by 2028 [1]. Also, among all the energy storage technologies, lithium ion based energy storage technology holds more share of the global market. As an effect of major policies and projects announced by several countries, such as low global warming potential equipment, low carbon emission, rural electrification, subsidy schemes to set-up infrastructure for EVs, etc, the battery market is projected to grow continuously.

Due to the exponential expansion of the EV market, the adoption of clean renewable energy sources, and the diversification of electrical systems, the demand for energy storage devices is rising day by day. The International Energy Association (IEA) estimates that, to align with net zero emissions by 2050, the world needs nearly about 600 GW of battery storage capacity by 2030 [2]. In 2020 globally, overall energy storage capacity was increased by about 5 GW with about an additional 40% hike in investment [2, 3]. Recent figures vary significantly, as the awareness of clean energy and demands for EVs is increasing. Indeed, some statistical reports predicted that the EV market may become larger than the present automobile market [4, 5]. However, with such a rapid scaling up in energy storage technology, LIBs cannot possibly address all of the applications due to the intrinsic limitations in their current form such as shorter cycle life, smaller power density, and high initial cost. Thus, the future development of LIBs with good cyclic behavior, higher power and energy density, safety, and easy recycling and optimum cost will be under serious consideration. The cost of LIBs is mainly from component materials, cell assembly, and packaging processes.

Wood *et al* performed extensive studies on the total cost breakdown for cell construction (excluding packing and assembly costs) by considering an average voltage of about 3.5 V and baseline of NMP/PVDF processing at standard electrode thicknesses. According to them, the cost of the electrode material is about 40.8%, electrolyte about 9.9%, separator about 26.0%, and current collectors about 6.1% of the total cost. In 40.8% electrode material cost, the cathode material contributes about 80%, which is about 32.7% of the total estimated cost of the cell construction [6]. As the LIB is a continuously upgraded and developing technology, R&D cost also matters a lot. In recent years, several advances have been made for cost reduction technologies both through continued development in novel component materials and automized battery assembly processes.

Among the various development strategies, research on electrode materials is major thrust area of serious concern. Considering the major share of electrode material cost in total battery cost, extensive work clearly needs to be done in the area of material development, optimization, and processing.

A number of nanomaterial classes for cathode electrode materials, including layered chalcogenides, layered transition metal oxides, metal phosphates, and silicates have the potential to offer LIBs in the future with improved electrochemical potential and low cost. However, there is still no universal electrode material that provides optimal battery performance for general application under all operating situations.

Thus, the comprehensive studies in the reported literature may shed some light on the future trends and challenges in LIB technology. The following sections in this chapter provide an idea about the key requirements of each component of LIBs and reviews of materials for cathodes with their current trends and future prospectives. Accordingly, in-depth analysis of the reported literature may shed some light on the potential developments and difficulties for LIB technology in the future.

2.2 Components and their role in LIBs

The four essential components of a lithium-ion battery cell are the cathode, anode, electrolyte, and separator. Each component's functions, material compositions, and typical materials are listed in table 2.1.

2.3 Interphases in Li-ion batteries

Interphase layers are formed in the initial charge–discharge cycles of LIBs due to the decomposition of electrode material in contact with the electrolyte. The uniqueness of the electrode/electrolyte interphases affects the chemical, electrochemical, and cyclic stability of the battery. In LIBs, two interphases are formed, one at the anode known as the solid electrolyte interphase (SEI) and another at the cathode known as the cathode electrolyte interphase (CEI). Both have equal importance from the perspective of material stability and electrochemical performance. If the anode has an electrochemical potential 'μ_A' higher than the LUMO of electrolyte, this may cause a reduction reaction at the anode if no SEI formed. Likewise, if the cathode has an electrochemical potential 'μ_c' lower than the HOMO of the electrolyte, it may cause an oxidation reaction at the cathode if no CEI formed. Thus, a proper combination of anode cathode and electrolyte is necessary for long life cyclic performance of LIBs. The formation of SEI and CEI is shown in figure 2.1.

2.3.1 Solid electrolyte interphase (SEI)

During the first charge–discharge cycle, the Li ion migrates from cathode to anode while charging and from anode to cathode during discharging. Due to decomposition of the electrolyte during these transfer processes through the electrolyte, a solid layer forms at the anode/electrolyte interface, known as SEI. It has a thickness of about 100–120 nm and is composed of a complex mixture of various organic and inorganic surface species, such as lithium oxide (Li_2O), lithium hydroxide (LiOH), lithium carbonate (Li_2CO_3), lithium fluorine (LiF), lithium alkyl carbonates ($ROCO_2Li$), etc. It is also known as a solid protective passivation layer which permits the transport of ions through it but avoids transfer of electrons (from the anode to the electrolyte), hence protecting the negative electrode. Additionally, it permits quick electrochemical reactions between the cathode, or positive electrode, and electrolyte, for improved LIB energy efficiency. However, a thick SEI above the threshold value might lead to LIB breakdown due to a sudden rise in internal impedance, which often occurs at high battery temperatures [7–9].

Table 2.1. Components and their functions in a lithium-ion battery.

Component	Function	Material
Current collectors	Provide mechanical support for the cathode and anode and act as current collecting electrodes by collecting and delivering electrons.	Al and Cu foil
Binder	Improves mechanical adhesion of cathode and anode material to the current collector	(PVDF or styrene butadiene rubber (SBR))
Anode	Gives lithium ions during discharging to the cathode	made up of active material, conductive additive and binder. A slurry of active material with binder (PVDF or styrene butadiene rubber (SBR)), carbon black as conductive material in (NMP) solvent coated on copper foil
Cathode	Gives lithium ions during charging to anode	made up of active material, conductive additive, and binder: various lithium containing compounds like LMO, LCO, LFP, NCA, NMC with PVDF as binder, carbon black as conducting additive in NMP solvent coated on aluminum foil
Electrolyte	Passes Li ions between the anode and cathode	Lithium salt: Lithium hexafluorophosphate ($LiPF_6$) Lithium perchlorate ($LiClO_4$) Lithium hexafluoroarsenate ($LiAsF_6$) Organic solvents: Ethyl methyl carbonate (EMC) Dimethyl carbonate (DMC) Diethyl carbonate (DEC) Propylene carbonate (PC) Ethylene carbonate (EC)
Separator	Prevents direct contact of electrodes and passes Li ions through its micro-porous permeable network to allow the active species to flow between the electrodes	micro-porous membrane of polyethylene or polypropylene such as: poly(vinylidene fluoridehexafluoro-propylene) (PVDF-HFP) poly(vinylidene fluoride-cochlorotrifluoroethylene) (PVDF-CTFE) poly(vinylidene fluoride tetrafluoro-ethylene (PVDF-TFE)

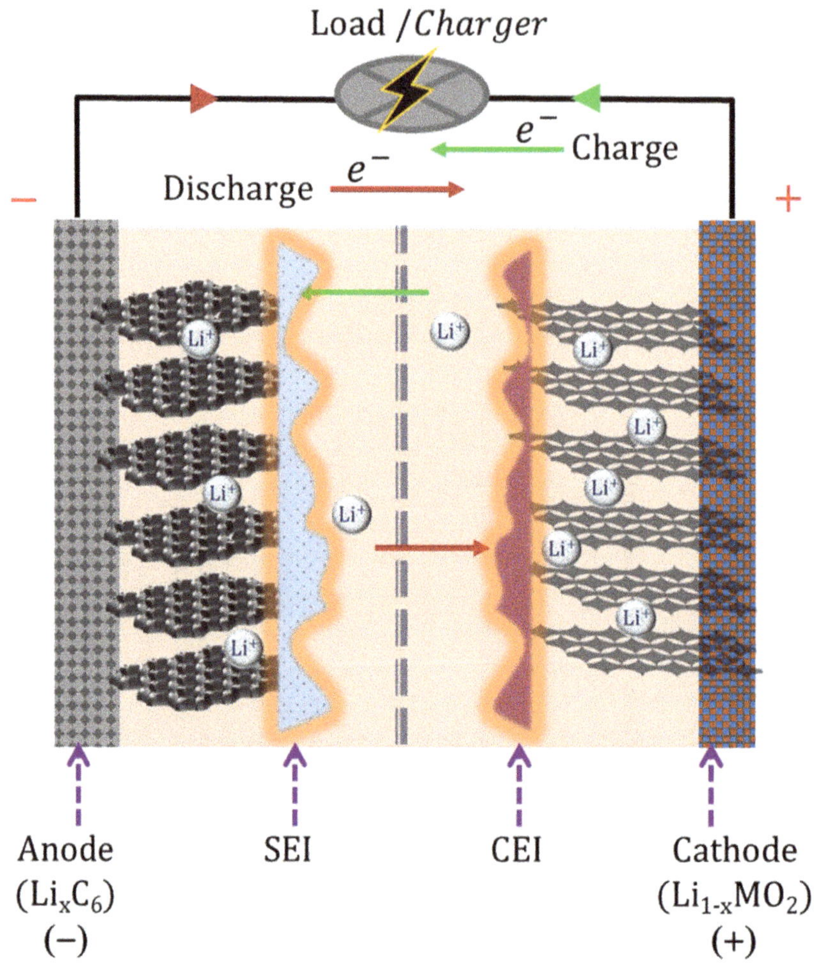

Figure 2.1. Formation of the SEI and CEI layers in a LIB cell configuration.

2.3.2 Cathode electrolyte interphase (CEI)

Similar to the SEI, the irreversible reaction at the cathode surface during initial charge–discharge cycles results in the dissolution of the electrolyte, creating a passivation layer while retaining the ionic transport channel, known as the CEI.

The CEI typically breaks down at high temperature, but it can spontaneously develop at high voltage over successive cycles. By incorporating additives into the electrolyte, CEI can also be synthesized artificially to be thick and stable. The electrochemical performance of the LIB can be improved by the stable CEI, which successfully supports the ion diffusion kinetics. Instead of being uniformly coated on the cathode electrode surface like the SEI layer, the CEI layer is heterogeneously dispersed/distributed, which could lead to capacity loss and a low retention rate [10–14].

2.4 Criteria for selection of the cathode material

The cathode, or positive electrode material, is an essential component of the LIB because it acts as an active source of Li ions during the charge and discharge process. In order to achieve good Li-ion diffusion kinetics, cyclic behavior, and high energy density, the cathode material needs to possess the following characteristics [15–17]:

- It should have electrochemical potential 'μ_c' higher than the HOMO of electrolyte;
- It should be abundant, low cost and environmentally friendly;
- It should be insoluble in the electrolyte and also noncorrosive;
- It should be the intercalation compound for incorporation of large amounts of lithium;
- It should have an ordered structure with reversibility during the intercalation and de-intercalation process;.
- It should have high reactivity with lithium and a transition metal cation with high oxidation state and multiple valencies;
- It should have good electrical conductivity and high lithium diffusion coefficient;
- It should have good mechanical, thermal, chemical, and structural stability;
- Its interconnected lattice network should provide shorter path lengths for fast Li-ion diffusion rates;
- It should be resistant to mechanical strain and structural distortion;
- Its structure should have an arrangement of Li ions and transition metal cations across separate layers for easy Li hopping;
- It should be processed easily as well as cheaply with several synthesis routes for uniform nanoparticle size and different surface morphology;
- It should ease the possibility of heteroatom doping or become a composite with another material, which may help to enhance conductivity and practical working potential.

2.5 Active cathode materials for LIBs

In Li-ion batteries the performance-governing and active source of Li ions is the cathode material. As the preferable positive electrode material is carbon based, which is not a source of Li, initially the priority for the Li active positive electrode is given. Spontaneous intercalation–de-intercalation of Li ions occurs during discharging and charging of the battery, respectively. The intercalation compounds are capable of insertion and extraction of lithium ions in a repeated manner into interstitial sites of its solid matrix without any substantial changes in the core structure. Thus, the intercalation reaction can be represented as:

$$Li^{+} + e^{-} + [ICH] \leftrightarrow \langle Li^{+}|ICH \rangle \quad (2.1)$$

where ICH stands for intercalation compound host. This repetitive reversible phenomenon is also referred to as the topotactic reaction. The intercalation capacity is limited by structural geometry, mass, and volume of the host material. The lithium-based intercalation compounds, owing to high electrochemical potential, electric conductivity, ionic conductivity, structural stability, and proficiency for shuttling Li ions through itself have been used as positive electrode materials in LIBs.

The cathode materials are divided into many classes according to their chemical structure, layered geometries, stoichiometry, and intercalation behavior. The following are the material classes that were used to categorize the cathode materials [18, 19].

- Layered dichalcogenides;
- Lithium layered metal oxides (e.g. $LiCoO_2$);
- Spinel:
    - Normal Spinel: manganese oxide (e.g. $LiMn_2O_4$), $LiMnCoO_4$, $LiMnCrO_4$, $LiMn_{1.5}Ni_{0.5}O_4$;
    - Inverse Spinel: $LiNiVO_4$, $LiCoVO_4$;
- Olivine: Polyanionic materials(e.g. $LiFePO_4$, $LiMnPO_4$, $LiNiPO_4$, $LiCoPO_4$), lithium cobalt phosphate ($LiCoPO_4$);
- Silicates: Polyoxyanion compounds (e.g. Li_2MSiO_4).

Figure 2.2 depicts the development of lithium-ion batteries since the introduction of lithium-metal batteries in 1972, using various cathode materials up until 2019.

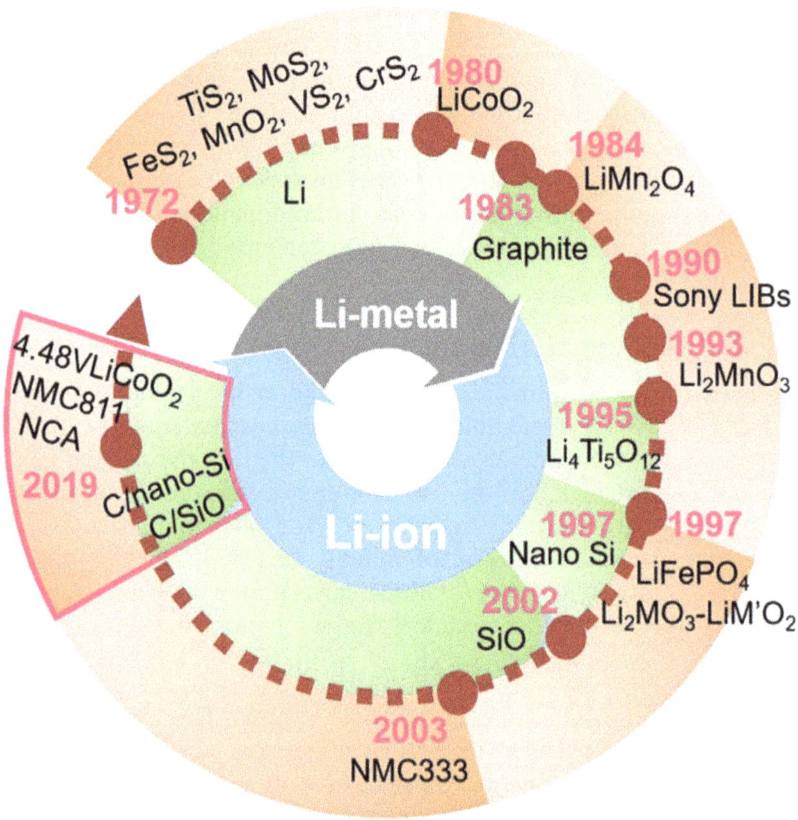

Figure 2.2. Development of lithium-ion batteries from lithium-metal batteries using various cathode materials between 1972 and 2019. Reproduced from [18] with permission from Elsevier.

The majority of layered transition metal chalcogenides are abundant enough, hence both are environmentally and ecologically favorable. Compared to oxides, they have better electrical conductivity, and their crystal structure is suited for intercalating lithium [16, 20]. The reversible intercalation of metal ions in layered metal chalcogenides was discovered in the early 1970s. Even the first cathode material for rechargeable LIBs was titanium disulfide (TiS_2). It has a layered structure with planes of Ti atoms sandwiched between two S layers. The layers are bound together in a hexagonal unit cell-like structure by weak Van der Waals forces. During discharging and charging, the Li-ion insertion or extraction does not cause any structural phase transition of TiS_2. This is due to the fact that interaction of Li ions in the host matrix establishes strong coulombic interaction with metal cations of the host, which resists structural distortion or expansion [21]. However, the *c*-axis lattice parameter was found to be increased due to interaction of Li ions [22]. Despite these advantages, it was not a good choice for future LIBs because of its limited potential and short cycle life [23]. Other metal sulfides such as FeS_2, VS_2, and MnS_2 have been explored, but they were also associated with the same problems, hence researchers started thinking about other classes of materials with similar structural geometry [24, 25]. Besides metal chalcogenides, layered transition metal oxides, lithium spinel compounds, polyanion-type compounds (olivine) and poly-oxyanion compounds (silicates) are receiving considerable attention as alternative cathodes for LIBs. Figure 2.3 shows the crystal structure of the most common layered, spinel, olivine, and silicate cathode materials.

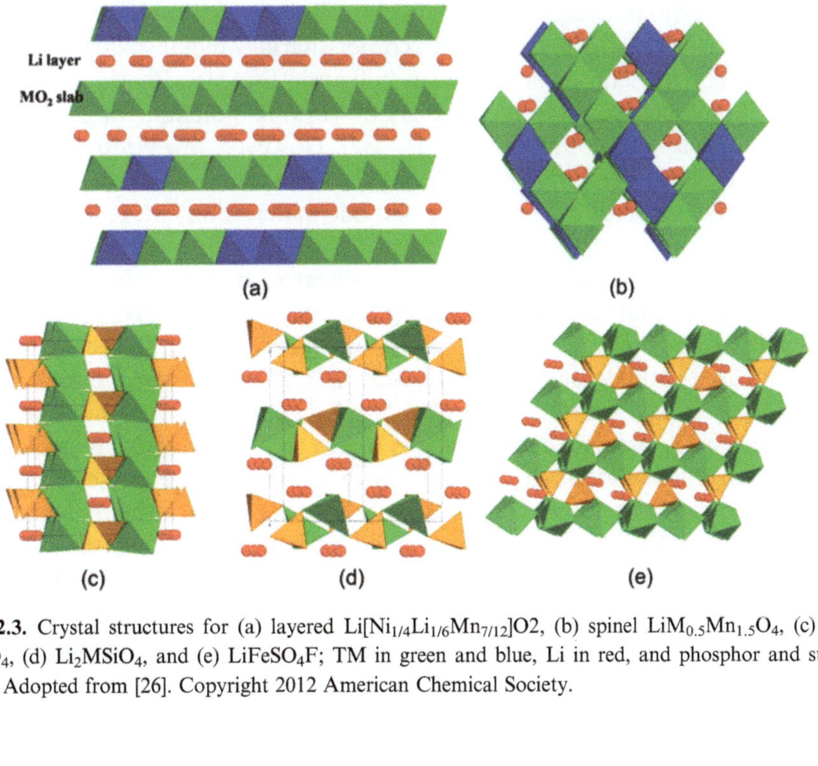

Figure 2.3. Crystal structures for (a) layered $Li[Ni_{1/4}Li_{1/6}Mn_{7/12}]O2$, (b) spinel $LiM_{0.5}Mn_{1.5}O_4$, (c) olivine $LiFePO_4$, (d) Li_2MSiO_4, and (e) $LiFeSO_4F$; TM in green and blue, Li in red, and phosphor and sulfur in orange. Adopted from [26]. Copyright 2012 American Chemical Society.

Due to their efficient Li-interaction mechanism, layered lithium metal oxides are the most extensively researched and commercially available cathode material for LIBs. As seen in figure 2.3(a), these materials have a layered chalcogenides-like structural geometry with metal cations sandwiched between oxygen atoms. They have a general formula $LiMO_2$, with a trigonal crystal structure. Here, M is a metal cation with an average +3 oxidation state, such as iron (Fe), nickel (Ni), cobalt (Co), manganese (Mn), or any binary combination of these [27]. In the structure Li^+ resides at octahedral (O), prismatic (P), or tetrahedral (T) sites based on the staking of MO_2 slabs [28]. Several layered metal oxides such as $LiCoO_2$, $LiNiO_2$, $LiMnO_2$ and their modified form by metal cation doping have been studied extensively.

The spinel family of electrode materials with the general formula AB_2O_4 showed high electrical conductivity, good structural stability, and environmentally-friendly properties [29]. The most common example of a spinel-type electrode material is $LiMnO_4$, in which the Li ions occupy tetrahedral sites, Mn atoms occupy octahedral 16d sites, and oxygen atoms occupy 32e sites as shown in figure 2.3(b) [30]. The Mn ion form a regular repeating framework with an edge-sharing MnO_6 octahedra providing intermediate tunnels for Li-ion insertion. However, the manganese (Mn) dissolution in electrolyte limits its performance [16, 20, 31].

The ordered olivine phosphate material family, often known as polyanionic compounds, are more environmentally friendly and generally safe compared to layered and spinel structures. They have a general formula $LiMPO_4$, where M is either Mn, Co, Ni, or Fe. The only difference between it and spinel is the nature of the two octahedral crystals, which are distinct and also of different size. The overall structure is orthorhombic composed of regularly arranged phosphate anion (PO_4^{3-}) tetrahedral units, which separate the corner-sharing metal oxide (MO_6) and lithium oxide (LiO_6) octahedral, as shown in figure 2.3(c) [16, 20, 31]. The electrode material with olivine structures offers good working potential of 3.4 V, and a theoretical capacity of about 170 mAh g^{-1} [32]. However, low electronic and ionic mobility influences the performance. Among the olivine cathode materials, $LiFePO_4$ has been widely explored as a promising cathode material for future EV technology [33, 34].

Polyoxyanion compounds, also known as silicates, are a new family of positive electrode materials that have gained interest for their high theoretical capacity, high energy and power density, cost-effectiveness, and safety issues. The general formula for silicate-based cathode materials is Li_2MSiO_4 (with M as cation, may be Fe and Mn). It can be formed with orthorhombic $Pmn2_1$, $Pmnb$, and monoclinic Pn, $P2_1/n$ phases. The $Pmn2_1$ orthorhombic phase is found to be thermodynamically stable. The structural framework comes with tetrahedral and octahedral sites occupied by Li, Mn, and silicon (Si), respectively. Li_2MnSiO_4 can achieve a theoretical capacity of 330 mAh g^{-1} with a working voltage of 4.0 V [35–37]. The main drawback with this kind of material is structural instability during operation due to Jahn–Teller distortion of the metal cations (Mn^{3+} ions in the case of Li_2MnSiO_4), leading to low specific capacity and poor cycle life [36–39]. Figure 2.4 shows the average discharge potentials and specific capacity achieved experimentally by some of the most common compounds from different classes of cathode materials. To date, several

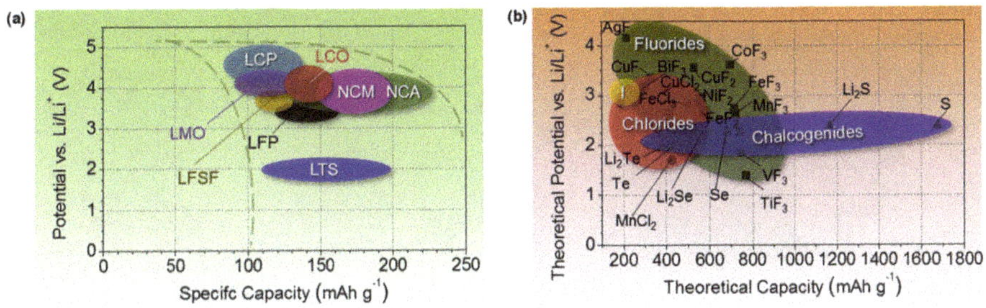

Figure 2.4. Average discharge potentials and specific capacity for different cathodes: (a) intercalation-type cathodes (experimental), and (b) conversion-type cathodes (theoretical). Reproduced from [20], copyright (2015) with permission from Elsevier.

materials like LMO, NCA, NMC, and LFP have been successfully adopted as cathodes for LIBs, mainly by the automobile industries for EV applications.

2.6 Ionic conductivity in the most common intercalation type cathodes

Great efforts have been made to understand Li-ion transport in layered intercalated compounds to obtain batteries with enhanced electrochemical performance. It is well known that energy and power density are significantly governed by Li-ion diffusion kinetics in the electrode materials. Several reports are available on the theoretical study of lithium diffusion using first-principles electronic calculation methods [38, 39]. It is well known that in most of the layered compounds the Li ions occupy tetrahedral sites between layers of octahedral oxygen atoms. The migration paths for lithium ions in different compounds are illustrated in figure 2.5 (b, d, and f)

The theoretical calculations showed that:

- In the case of layered lithium metal oxides the migration or diffusion of octahedrally coordinated Li^+ ions takes place from one octahedral site to another vacant octahedral site through the intermediate tetrahedral site of oxygen;
- In the spinel family of electrode materials, tetrahedrally coordinated Li ions migrated between neighboring tetrahedral sites via the octahedral site;
- In an ordered olivine phosphate structure the Li ion can transport between nearby Li sites along the [010] direction (*b*-axis) following a curved oct-tet-oct trajectory.

2.7 Recent advances in active cathode materials

This section explores the fundamental studies and advancements in the development of cathode materials for LIBs. Efforts to improve the electrochemical performance, scientific challenges and approaches to overcome the challenges are also discussed.

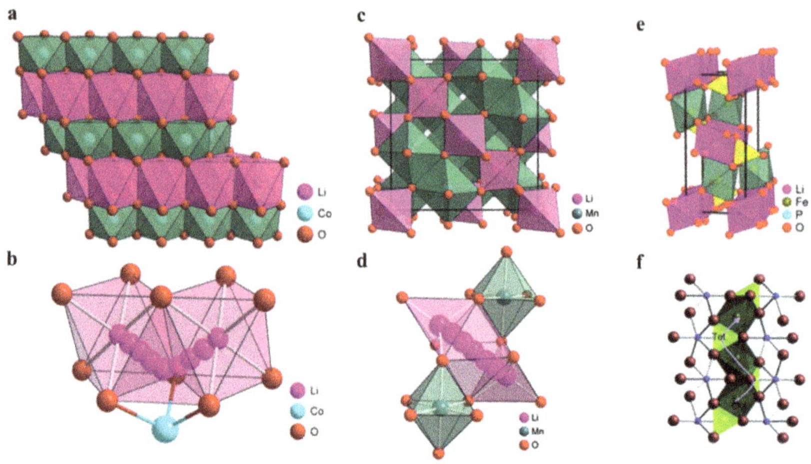

Figure 2.5. Crystal structure and diffusion pathways for Li ions in (a and b) layered $LiCoO_2$, and (c and d) spinel $LiMn_2O_4$. Reproduced from [38], copyright (2013) with permission from Elsevier. (e and f) Olivine $LiFePO_4$. Reprinted from [39]. Copyright 2009 with permission of Royal Society of Chemistry.

2.7.1 $LiCoO_2$ (LCO)-based positive electrode material

Since its introduction in 1980 by *Prof. John B Goodenough*, $LiCoO_2$ has emerged as the most leading commercialized material in LIBs, as it has moderate structural stability during insertion and re-insertion of ions, high operation voltage, high energy density, good capacity retention, and excellent cycle life [40–42]. It has a redox potential of 4 eV and also shows relative fast kinetics for both ion and electrons. Later on, in 1991, the Sony company demonstrated the first successful lithium-ion battery with LCO as the cathode and graphite anode in $LiPF_6$ in propylene carbonate (PC) electrolyte [43]. It was predicted that, as a lithium-ion intercalation material, LCO would show reversible de-intercalation and re-intercalation properties up to a potential approximately 4.2 V vs. Li^+/Li [43, 44].

As shown in figure 2.3, in layered lithium oxides, the Li ion resides at the octahedral site surrounded by tetrahedral transition metal cations, which are located in between two oxygen atoms. Thus, the transition metal cation layers are separated from the Li-ion layers by oxygen atoms. The LCO has an α-$NaFeO_2$ cubic rock-salt structure with $a = 0.2816$ nm, $c = 1.4056$ nm, and a ratio of c/a about 4.899. Ideally, the Li^+ and Co^{3+} ions are present with a closely packed network of oxygen atoms. The difference in ionic radii between the Li^+ and Co^{3+} ions (0.70 vs. 0.52 Å), and in the nature of the bonds (ionic for Li–O and covalent for Co–O) induces an ordering of the cations in successive planes. The framework crystallizes in the rhombohedral system (space group R-$3m$) and consists of an alternated stacking of CoO_2 slabs and LiO_2 interslab spaces made up, respectively, of CoO_6 and LiO_6 octahedra. This oxide is thermodynamically stable only in its lithiated state and is an electronic insulator in its stoichiometric form $LiCoO_2$. The band gap of LCO is about 1.7–2.7 eV. Due to difference in ionic radii, Li^+ and Co^{3+} can have dissimilar interaction with oxygen, because of which the closely packed network of oxygen atoms get deviated and showed trisymmetry with space group *R*-3m. The edge-shared CoO_6 octahedral arrangement is responsible for its good electronic conductivity. Insertion/extraction of the Li^+ ion occurs between the strongly bonded CoO_2 layers. The migration or diffusion of the Li^+ ion takes place from one CoO_6 octahedral site to another vacant octahedral site through the intermediate tetrahedral site of oxygen [44].

To achieve high energy density, good capacity retention, and excellent cycle life with LCO, several physical and chemical synthesis methods have been adopted for synthesis of LCO such as pulsed laser deposition, radio-frequency (RF), magnetron sputtering, chemical vapor deposition, sol–gel, reflux, hydrothermal, solution combustion, microwave, ultrasonic radiation, thermal decomposition, electrostatic or electro spinning spray deposition (ESD), template assisted synthesis, molten salt process, etc. Massive efforts have been taken to enhance the performance of LIBs using LCO. Most of the studies were carried out regarding structural deterioration, chemical and phase evolution at the surface of the electrode [45–61].

The most common route for synthesis of LCO is by high temperature solid-state reaction. At high temperature, the lithium salt is mixed with cobalt oxide under a long reaction time. Sometimes a binder such as polyvinyl alcohol (PVA) can be added to avoid dissolution of the fine particles in the intermediates. However, because of inadequate mixing of precursor salts at a lesser period, the reaction and evaporation of lithium ingredients and coarsening of the product at prolonged calcination in solid-state reaction resulted in low capacity retention and shorter cycle life.

The liquid assisted method does not require high temperature treatment and also helps to tune the stoichiometry of LCO easily, because of subatomic-level reactions between almost all Li^+ and Co^{3+} ions. The electrochemical properties can be enhanced by optimizing processing parameters so as to prepare high quality $LiCoO_2$ with nanosize and different nanostructures. Nevertheless, the crystallinity of nano-sized products in most of the liquid assisted methods is not much higher as compared to those prepared at high temperatures, thus further high temperature treatment is necessary. Also, there is a chance of intermediate phase formation which results in poor electrochemical performance [62].

The sol–gel method is also one of the economic, low-temperature chemical synthesis methods used to prepare LCO with better crystallinity and uniform particle distribution as compared to the solid-state method. During preparation, the pH of cobalt salt solution was adjusted by a complexing agent like ammonia to form a gel. A thin-filmed electrode can be obtained either by spin coating (in the case of gel) or by the ink-jet printing method (in the case of powder). The particle size can be controlled by adding some acids which serve as carriers for ions. The most common additive acids are citric acid, oxalic acid, acrylic acid, polyacrylic acid (PAA), succinic acid, and tartaric acid [51, 52]. The layered structure with various morphologies can be prepared using a hydrothermal method. For LCO, salts like LiOH, $Co(NO_3)_2$, and H_2O_2 are mixed and react at temperatures in the range of 120–250 °C for 1–36 h in a liner under controlled pressure. Particle sizes in the range of 70–200 nm can be prepared with good crystallinity without post-heat treatment [54]. High-purity single-phase LCO can be prepared by electro spinning spray deposition at high temperature in the range of 600–800 °C. It is reported that the post-annealing treatment improves the electrochemical performance of the battery due to enhancement in the crystallinity of LCO. Interestingly, it is found in the case of the physical method that Li diffusion planes vertical to the substrate surface in a prepared electrode facilitate the easy intercalation–de-intercalation of Li ions compared to planes parallel to the substrate surface, which may slow Li-ion intercalation–de-intercalation [59].

The predicted theoretical capacity of LCO is around 274 mAh g^{-1} (considering all Li is removed electrochemically); however, the constraint over removal of Li^+ from LCO due to structural instability, limited it to only 50% of the theoretical value specific capacity, corresponding to ≈140 mAh g^{-1}. Thus, for reversible de-intercalation and re-intercalation the cut-off voltage should be limited to 4.2 V vs. Li^+/Li [63].

The synthesis conditions and Li/Co molar ratio influence the LCO structure and thus strongly affect the electrochemical performance. Reports in the literature show that desired electrochemical performance of LIB can be achieved by using over-stoichiometric LCO as the cathode material. For instance, theoretical reversible capacity for $Li_{1-x}CoO_2$ with $0 \leqslant x \leqslant 0.5$ is about 156 mAh g^{-1} with a plateau at 4 V (vs. Li^+/Li) in a typical discharge curve [51–55].

The predicted highest mole of Li^+ ions extracted or inserted per formula unit during reversible de-intercalation and intercalation in $Li_{1-x}CoO_2$ is 0.5. The $Li_{1-x}CoO_2$ with $x = 0.5$ undergoes reversible phase transition from trigonal into monoclinic symmetry during the charge–discharge process. However, further de-lithiation, i.e. for $x > 0.5$, results in an unstable structure $Li_{1-x}CoO_2$, leading to mechanical damage, decomposition, and dissolution of cobalt in organic electrolytes, and oxygen loss. This affected the electrochemical performance severely [16, 64–66].

It has been reported that the increase in cut-off voltage beyond 4.55 V initiated a structural change in LCO and dissolution of cobalt in organic electrolytes. This resulted from irreversible phase transition as an impact of large anisotropic structural change. A large change in ratio of c/a causes a transition from the hexagonal to monoclinic phase, resulting in structural damage and hence capacity fading [67]. Thus, LCO with high crystallinity and structural stability during reversible lithium intercalation and de-intercalation is vital to maintain good electrochemical performance of the battery [43, 44]. The nanostructured LCO has high crystallinity, and is expected to provide high capacity. Tang *et al* reported the electrochemical performance of nanosized LCO at different current densities in Li_2SO_4 aqueous electrolyte. They reported an initial discharge capacity of 143 mAh g^{-1} at 7 C between 0 and 1.05 V. The cycling performance was attributed to shortening the diffusion path in nanoparticles for the migration of the Li ion [68].

Azib *et al* synthesized $LiCoO_2$ thin films by an hydrothermal route using water and ethanol as solvent. The effect of the synthesis parameters like solvent ratio, pressure, reaction time, and temperature on the structural and electrochemical properties of LCO was evaluated. The prepared films exhibited a hexagonal phase of

LCO and showed a specific capacity of 120 mAh g^{-1} with better capacity retention for the sample deposited at 125 °C and annealed at 400 °C [69]. Duffiet *et al* studied the effect of initial Li/Co stoichiometry in $LiCoO_2$ on the electrochemical performance of LIBs by *in situ* X-ray diffraction. It was reported that the phase transition at high voltage occurred for $1.00 \leqslant Li/Co \leqslant 1.05$ [70]. Jiang *et al* conducted systematic investigations on the electrochemical performance of LCO at a potential over 4.7 V. They reported specific capacity from 160 to 180 mAh g^{-1} in different voltage windows in the range of 2.7–4.7 V. The authors reported capacity retentions of 88% for 2.7–4.2 V cycling, and 19% for 2.7–4.7 V cycling. Low coulombic efficiencies at 2.7–4.7 V cycling indicated that, during the discharge process, lithium ions are not able to be re-inserted back significantly. According to the authors, high voltage cycling caused bulk structure degradation, which was further responsible for capacity decay over 4.7 V [71]. Liang *et al* synthesized LCO at lower temperatures (650 °C and 750 °C) via the solid-state reaction method. The effect of Li/Co molar ratio on the physicochemical and electrochemical properties was examined. The $LiCoO_2$ prepared at 750 °C and 2 h with a molar ratio (9:1) between $LiOH{\cdot}H_2O$ and Li_2CO_3 showed a discharge capacity of 98.3 mAh g^{-1} and 80.7% capacity retention after 50 cycles at 1 C [72]. Reddy *et al* synthesized lithium cobalt oxide using a template-assisted sol–gel synthesis route using cherry blossom leaf (CBL) templates followed by calcination at 800 °C for 12 h. A higher initial discharge capacity of 166 mAh g^{-1} and 81% capacity retention have been reported for a sample prepared with cherry blossom leaf templates. The pristine LCO exhibited an initial discharge capacity of 127 mAh g^{-1} and 72% capacity retention at the cut-off potential of 2.5~4.2 V vs. Li^+/Li. The higher discharge capacity, about 30.7% in LCO-CBL attributed to the enhanced surface area, increased about 79.7% more than pristine LCO [73].

Thus, the structural phase transformation, surface structure decay, decomposition of electrolyte, inhomogeneous reaction mechanism, O_2 loss, Co dissolution, and low Li-ion diffusivity due to a decrease in Li in the LCO electrode over 4.2 V significantly leads to capacity fading and less cycle life. Nevertheless, the presence of expensive and toxic cobalt in LCO leads to an increase in overall cost and safety measures. One of the major drawbacks with LCO is its thermal stability and low reversible capacity, limited to 135 mAh g^{-1} with 0.5 exchange of Li^+ per mole [16, 20, 31, 74].

The phase transformation and surface degradation resulted in non-uniform distribution of the state of charge at the different parts of the electrode material, which leads to a decrease in cyclic performance. The inhomogeneous reaction at the different particles in the electrode material causes cracks, particle pulverization, and loss of ohmic contact between the current collector and electrode material. This leads to poor cycling performance and hence prevents massive industrialization of the material for energy storage application [75]. Some of the reported results suggest that the specific capacity loss due to the phase transition from hexagonal to monoclinic phase of LCO happened mostly as a result of an increase in surface impedance, which can be controlled by surface treatments [76].

To overcome these issues and provide long-cycling stability to LIBs with LCO at high operating potential, several strategies have been explored such as surface coating, nanocomposites, and doping with transition metal cations. Also, to obtain more energy density, specific capacity, and cycling stability over 4.5 V, the surface structural stability of LCO needed to be improved. Gu *et al* have investigated the electrochemical performance of LIBs with bare LCO and $LiCo_xMn_{2-x}O_4$ coated LCO layer (LCMO-LCO) as cathode electrodes in a 1 M $LiPF_6$ electrolyte. The study reported discharge capacities of 104.9 mAh g^{-1} and 143.8 mAh g^{-1}, with retentions of about 67.4% and 66.1%, after 300 cycles for bare and LCMO-LCO, respectively. The enhanced cycle performance was attributed to the surface structure stability, cathode electrolyte interface stability, and high conductivity [77]. The change in chemical composition of LCO, i.e. substituting Co by different heteroatom dopants such as Ti, Ni, Mn, Cu, Al, Mg, Mo, Cr, W, Fe, Zr, and Sb, showed enhanced electrochemical performance at higher discharge current densities [68, 78–86].

The doping not only suppresses the phase transformations but also increases the interlayer spacing to ease the Li^+ diffusivity. Some of the elemental doping was found to improve the electronic conductivity, structure evolution, and operating potential too. For instance, Sun *et al* studied the electrochemical performance of Ti-doped LCO (Ti-LCO) at a high cut-off voltage of 4.5 V without structural damage. No change in Li^+ diffusion coefficient occurred even though phase transition takes place. Ti-LCO exhibited high capacity of 205 mAh g^{-1} at 0.1 C with 97% retention after 200 cycles [67]. Kim *et al* studied Ti-LCO by a first-principles simulation method and thermodynamic investigation. The small amount of Ti substitution (<0.15%) in the LCO lattice was expected to be the most possible form of doping. Most probably, the formation of Ti-containing secondary phases, such as Li_4TiO_4, Li_2TiO_3, or TiO_2 was formed rather than the incorporation of Ti in the LCO lattice. Still, the LIB based on Ti-LCO showed enhanced cyclic stability at high voltage due to an increase in conductivity [78]. It was reported that a small amount of Ni doping in LCO (Ni-LCO) enhances the structural stability and decreases charge transfer impedance, resulting in better cyclic performance. However, more Ni in Ni-LCO blocked the lithium diffusion path, increased charge transfer impedance, and hence resulted in reduced reversible capacity and coulombic efficiency of LCO [79]. Kim *et al* prepared Cu-doped LCO by a facile ion-exchange process at low temperature. The egg-shell structured Cu-doped LCO demonstrated good cyclability and high coulombic efficiency with a discharge capacity of 133.8 mAh g^{-1} and capacity retention of 90% at 20 C over 500 cycles [81]. Er-Rami *et al* investigated the effect of Al doping in LCO on electrochemical properties. The better cyclic performance at about 4.6 V vs. Li^+/Li was attributed to suppressed cobalt dissolution due to effective Al doping. The TEM analysis of the electrode revealed that Al doping causes the formation of a stable CEI layer, resulting in improved Li-ion diffusion and electrical conductivity [82]. Mg is also preferred as one of the best choices for doping in LCO because of its high melting point, good electrical conductivity, and good oxidation resistance. The doping of Mg in LCO leads to an improvement in electrical conductivity and hence electrochemical performance [83]. Nowadays, multiple doping strategies have emerged as a promising approach to suppress the individual limitations of dopants. The doping of Mg can enhance the electrical conductivity while with Al the structural stability can be improved by suppressing Co dissolution. Ni doping can suppress phase transition due to the pillar effect. Doping nickel and manganese in various combinations to replace cobalt leads to higher energy density at lower cost [86–91].

Zhang *et al* [92] reported the improved electrochemical performance of LIBs with Ti–Mg–Al co-doped LCO (TMA-LCO) as the cathode, prepared by a solid-state reaction method at 900 °C.

The electrochemical performances of electrodes were tested in a half-cell (LCO/Li half cell) assembly using coin cells (CR2032) and full-cell assembly with LCO and graphite. A comparison of the cycle performances of half cells with bare LCO and TMA-LCO is shown in figure 2.6(a). At the high-charging cut-off voltage of 4.6 V, TMA-LCO exhibits better cycle stability in half cells compared to bare LCO. The half cell with TMA-LCO exhibited reversible discharge capacity of 174 mAh g^{-1} at 0.5 C and capacity retention of 86% over 100 cycles. Figure 2.6((b) and (c)) shows the charge–discharge profiles of bare LCO and TMA-LCO half cells for the 1st, 5th, 10th, 50th, and 100th cycles. After 50 cycles, it can be seen that the voltage profiles of bare LCO have greatly degraded, showing that the structural deterioration of bare LCO is more severe than that of TMA-LCO.

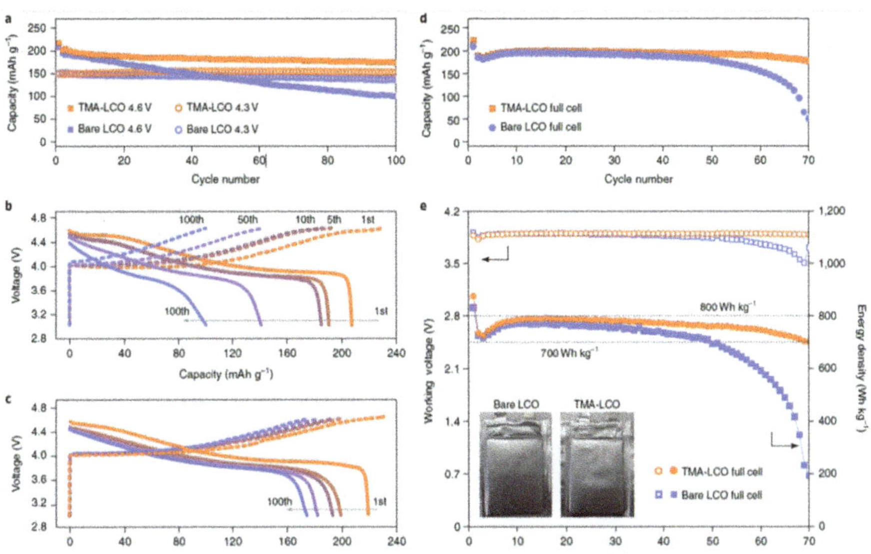

Figure 2.6. Electrochemical performance of LIBs with LCO and Ti–Mg–Al co-doped LCO (TMA-LCO) cathode materials. (a) Comparison of cycle performances of half cells with bare LCO and TMA-LCO. Charge–discharge profiles of half cells for the 1st, 5th, 10th, 50th, and 100th cycles with (b) bare LCO, (c) TMA-LCO. (d) Cyclic performance of full pouch cells with bare LCO and TMA-LCO, and (e) full-cell discharge voltage, and the cathode material's energy density in relation to cycle number. (Inset: pouch cells after the 70th cycle.) Reproduced from [92], copyright 2019 The Author(s), under exclusive licence to Springer Nature Limited, with permission of Springer.

Figure 2.7 compares the cycle and rate performances of Ti, Mg, and Al single-element-doped $LiCoO_2$ in comparison to those of bare LCO and TMA-LCO. As can be seen, single-element doped LCO performed better than unmodified LCO but worse than TMA-LCO, indicating that the synergistic effects of codoping improved cyclic performance. Figure 2.6(d) illustrates the cyclic performance of full pouch cells in the voltage range of 3.0–4.55 V (equivalent to 4.6 V vs. Li/Li^+) at room temperature with bare LCO or TMA-LCO cathodes and commercial graphite anodes. As can be seen, the capacity of bare LCO rapidly decreases to 51.3 mAh g^{-1} after 70 cycles, whereas the TMA-LCO cell retains better capacity of 178.2 mAh g^{-1} after the same number of cycles. The cycle performance of the bare LCO full cell has been greatly diminished as a result of irreversible structural distortions. This is further demonstrated by the obvious gas generation in the cycled pouch cell, as seen in the inset of figure 2.6(e). The enhanced performance of TMA-LCO in both the

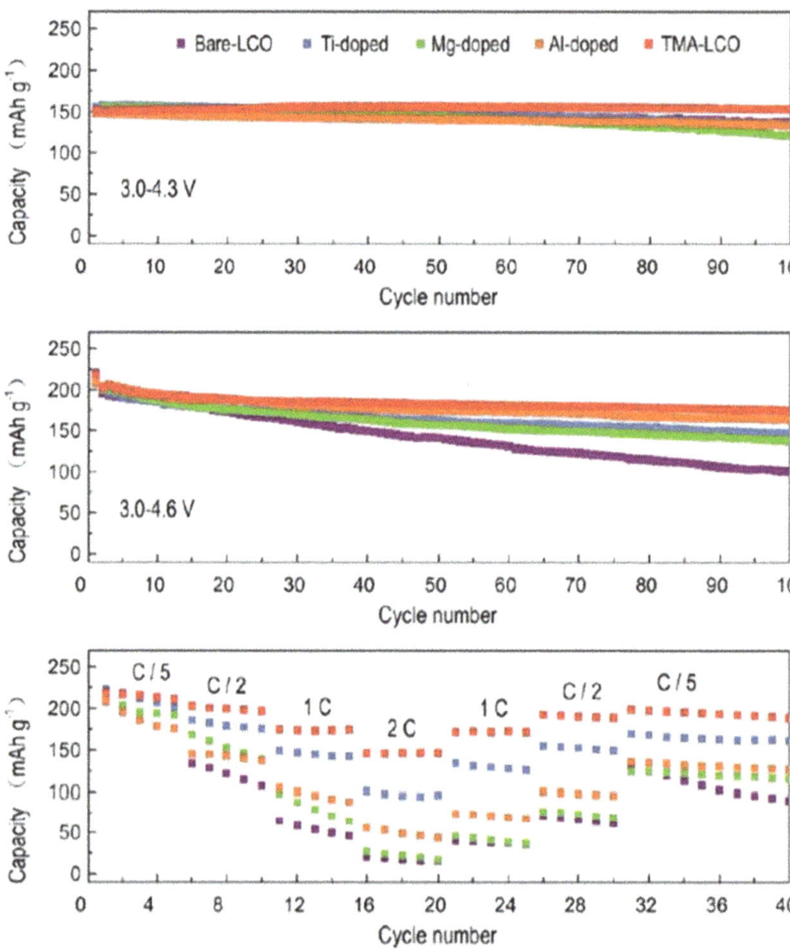

Figure 2.7. Comparison of the cycle and rate performances of Ti, Mg, and Al-doped LCO, bare LCO, and TMA-LCO. Reproduced from [92], copyright 2019 The Author(s), under exclusive licence to Springer Nature Limited, with permission of Springer.

half cell and the full cell at a high-charging cut-off voltage was attributed to the synergistic effect of enhanced electrical conductivity, and the modified microstructure favorable for ion diffusion kinetics.

Along with doping, coating of LCO with other conducting materials was found to be a successful strategy for enhancement of the cyclic stability with good electrochemical performance due to the change in surface composition. Several classes of materials such as metal oxides, fluorides, phosphates, and carbonates have been used for effective surface coatings of LCO. Among these, the performance of metal oxides coated with LCO showed noteworthy electrochemical performance due to an improvement in electrical conductivity and ion diffusion kinetics [60].

Shen *et al* reported surface modification of LCO by lithium ionic conductor $LiAlSiO_4$ coating. Improvement in the electrochemical performance of coated LCO at the high cut-off voltage of 4.55 V was attributed to the interface and structural stability as well as improved Li diffusion rate in LCO. The surface-modified LCO electrode showed a reversible capacity of 209.3 mAh g^{-1} with 89.3% capacity retention after 100 cycles at 4.55 V [93].

2.7.2 $LiNiO_2$ (LNO)-based positive electrode material

To replace Co in LCO, other transition metals (Mn, Ni, and Fe) based layered lithium oxides have been explored extensively. Among these, lithium nickel oxide ($LiNiO_2$, LNO) emerged as an alternative to LCO because of its high theoretical discharge capacity of 275 mAh g^{-1} and relatively lower costs as Ni is abundant, less toxic, and hence is cheaper [94, 95].

LNO has a similar structure to LCO, but the strong Ni–O–Ni covalent bonding in LNO results in semiconducting properties [95].

In LNO, there is no overlap of e_g bands of trivalent nickel and 2p band of oxygen, which leads to high Li-ion diffusivity without any extraction of oxygen from the host lattice. However, the lithium-ion diffusion is greater at higher temperature due to favorable ion hopping with a sufficient activation energy [95, 96]; whereas preparation of LNO with stoichiometry is found to be a difficult task due to the close ionic radii of Ni^{2+} (0.69 Å) and Li^+ (0.76 Å). It has been reported that LNO often occurs in $Li_{1-x}Ni_{1+x}O_2$ forming with some Ni-ion occupying sites in Li layers [95].

The particle size and distribution also showed impact over the electrochemical performance of LCO. Smaller-sized particles resulted in high capacity but low stability, while large particle size is favorable for thermal stability, but enhances the charge transfer. Kong *et al* theoretically studied the effects of non-stoichiometric Ni-rich LNO on electrochemical performance using density functional theory. It was predicted that the excessive Ni occupying the Li sites created a 'pillar effect' that increases the structural stability. The extra Ni-defects were responsible for the charge disproportionation reaction that reduces the Jahn–Teller distortion of Ni^{3+} ions. However, the Li depletion regions surrounding extra Ni-defects act as hurdles for Li-ion diffusion, which resulted in capacity fading and poor cyclic performance [97].

Moreover, extra de-lithiation causes oxidation of Ni ions, changing their state from Ni^{2+} to Ni^{3+} in the Li layer. This leads to a decrease in the interslab distance, limiting the diffusion of Li^+ ions, which resulted in a significant decrease in irreversible capacity. The non-stoichiometric Ni-rich LNO causes formation of the two-phase hexagonal and O1 phase as a result of change in the interslab distance [96, 97].

Various physical and chemical synthesis techniques have been reported for LNO such as plasma sputtering, combustion method, solid state, redox reaction in solution, emulsion method, sol–gel, and co-precipitation [98–101]. During synthesis, too high a temperature leads to impurity phases of other oxides along with LNO, while too low a temperature results in poor crystallinity [102]. The most commonly employed method for the synthesis of LNO is solid-state synthesis in which a mixture of hydrated lithium hydroxide ($LiOH{\cdot}H_2O$) and nickel (II) hydroxide ($Ni(OH)_2$) is heated at a high temperature of about 900 °C. It was observed that the sample prepared in an oxygen atmosphere at 700 °C exhibited higher discharge capacity of about 200 mAh g^{-1} in a 3.0–4.3 V voltage range. The moderate amount of Ni occupying the Li sites is favorable for the electrochemical performance of LNO. Deficient or excess Ni causes structural instability and hence poor cyclic performance [99, 101].

Improved cyclic stability and reduced capacity fading rate can be achieved by doping or partial substitutions for nickel in LNO. Several heteroatoms such as Mn, Co, Li, Al, F, Mg, Ti, Fe, Cu, Zn, Ga, and Nb have been investigated for doping into LNO to improve structural stability and enhance electrochemical performance. The substitution causes better cationic distribution that enables the structure to be more stable [103]. For instance, a small amount of Mg doping in LNO substituted Ni^{2+}, while a large amount occupies the Li sites. In both cases an enhanced performance was observed. Doping of Al in LNO prevented it from over charge because of its inert behavior and also enhanced the Li-ion diffusivity [104]. It has been observed that the partial substitution of Ni by Mn in LNO prevents structural desorption and phase transitions. The partial substitution of Ni by Co resulted in enhanced reversible capacity of about 180 mAh g^{-1}. The enhanced performance was attributed to structural stabilization, as both Ni and Co exist in the same period with almost similar extra-nuclear electron distribution, and the radius of Co^{3+} is less than that of Ni^{2+} [104]. The cyclic stability of LNO can be enhanced by partial substitution of aluminum for nickel, which endorsed the improvement in thermal stability of the electrode considerably [103].

2.7.3 $LiNi_{0.8}Co_{0.15}Al_{0.05}O_2$, (NCA) based positive electrode material

Nowadays, lithium nickel cobalt aluminum oxide ($LiNi_{0.8}Co_{0.15}Al_{0.05}O_2$, NCA) is emerging as a promising candidate for future LIBs, as doping of Co results in enhanced reversible capacity and Al substitution shows an improvement in thermal stability. The incorporation of Al is also found to enhance structural stability and increase in operating potential as a result of a strong Al–O bond compared to the Ni–O bond. The substitution of trivalent Co and Al in LNO at the Ni sites hinders the formation of Ni^{2+} and the cation mixing, resulting in a stabilized crystal structure during repeated de-insertion and re-insertion of Li^+. The more electrochemical active nature of Co^{3+} improved the electrochemical performance. The optimized performance was reported with Co (15 mol%) and Al (5 mol%) with high discharge capacity of about 200 mAh g^{-1} and excellent cycling stability even at high temperature (60 °C) [105, 106]. Considering such progress, in the upcoming years the energy density for NCA is expected to reach about 300 Wh kg^{-1}, and hence will stand as a promising candidate for future EV technology [107].

2.7.4 $LiMnO_2$ (LMO) based positive electrode material

After LCO, lithium manganese oxide ($LiMnO_2$) has emerged as a promising layered structure as an economic and environmentally-friendly electrode material due to its high theoretical capacity, non-toxicity, and low cost of MnO_2. LMO is a well-commercialized material and has about 80% share in batteries used for memory backup devices, watches, and calculators. It comes with two distinct structures, either rhombohedral or orthorhombic. Orthorhombic $LiMnO_2$ (o-$LiMnO_2$) is considered as a potential cathode material and is found to be thermodynamically more stable than the rhombohedral structure, also showing high capacity (285 mAh g^{-1}) and high potential vs. lithium [108, 109].

However, transformation of both orthorhombic and rhombohedral LMO to a spinel structure was observed during the cycling process, which resulted in a drop in voltage profile. The decrease in performance was also attributed to closer ionic radii of Mn^{3+} and Li^+, Mn dissolution, poor crystallinity, and Jahn–Teller effect of high spin Mn^{3+} ions [110, 111].

Several efforts have been made in order to improve the structural stability and cyclic performance of LMO, which include cation doping, transition metal oxide coating, and synthesis of different nanostructures. The doping and coating strategy resulted in better performance due to prevention of Mn metal dissolution, reduction of surface reaction, and Jahn–Teller distortion-induced lattice strain. Cations like aluminum (Al), boron (B), chromium (Cr), cobalt (Co), nickel (Ni), molybdenum (Mo), and titanium (Ti), were used for doping in LMO [112].

Xiao *et al* synthesized orthorhombic nanoparticles and nanorods of $LiMnO_2$ by hydrothermal methods. The LMO nanorods exhibited higher discharge capacity of about 200 mAh g^{-1} and improved cyclability. After 30 cycles the nanoparticles showed a greater drop in capacity from 170 mAh g^{-1} to 110 mAh g^{-1}, whereas in the case of nanorods it decreases less, from 200 mAh g^{-1} to 180 mAh g^{-1}. The enhanced electrochemical performance of LMO with nanorods was attributed to structural stability and ease of the path for charge transport through the nanorods. According to the study, a long potential plateau at about 3.0 V (vs. Li^+/Li) in the discharge curve is characteristic of o-$LiMnO_2$, while a plateau of about 4.0 V indicates transformation of the structure from orthorhombic into a spinel phase [113]. Tong *et al* prepared mesoporous orthorhombic $LiMnO_2$, which exhibited maximum discharge capacity of 191.5 mAh g^{-1} and a high retention of about 84.9% after 50 cycles at 0.1 C-rate [114]. Ji *et al* prepared rod-like orthorhombic $LiMnO_2$ nanocrystals using a temperature-controlled microwave hydrothermal route, which offers synthesis of nanomaterials at low temperature and in short time. The electrochemical performance exhibited by the as-synthesized o-$LiMnO_2$ nanorods showed a maximum discharge capacity of 194.2 mAh g^{-1} at 0.1 C-rate after several charge–discharge cycles between 2.2 V and 4.4 V vs. Li^+/Li at room temperature [115]. Xu *et al* synthesized microcubes of $LiMnO_2$ by the hydrothermal method. The microcubes with interlaced nanoplates showed a specific capacity of 197 mAh g^{-1} and 134 mAh g^{-1} at 0.1 and 1 C, respectively, and retention of about 93% after 50 cycles. The enhanced performance was attributed to the ease of transport of lithium ions due to a shortened path length and sufficient electrode–electrolyte contact, which led to more active electrochemical sites [116].

Recently, some studies on a composite of orthorhombic $LiMnO_2$ were also carried out for further improvement in specific capacity and stability. Shen *et al* prepared conductive o-$LiMnO_2$/carbon nanotubes (CNTs) composite frameworks via an ingenious one-step dynamic hydrothermal method. The electrochemical performance was monitored along with CNT content in o-$LiMnO_2$/CNT composites. Excellent electrochemical performance was achieved with 5 wt% CNTs o-$LiMnO_2$/CNT composites with a specific capacity of 204.9 mAh g^{-1} and high capacity retention of 93.7% after 50 cycles at 0.1 C. The authors stated that the enhanced performance was due to the significant improvement in lithium diffusion coefficient measured using the galvanostatic intermittent titration technique (GITT) [117]. Figure 2.8 displays SEM micrographs for pristine o-LMO and LMO/CNT composites. As shown in the inset of figure 2.8(a), pristine o-LMO is composed of a porous network of nanoparticles and no effective change in it was observed even after the addition of CNT (figure 2.8(b)).

Figure 2.9 shows cyclic voltammetry (CV) of pristine o-LMO and LMO/CNT composites within the potential range of 2.0–4.5 V. All samples exhibited a quasi-rectangular curve with an intense oxidation peak at 3.5–4.1 V and a comparatively less intense reduction peak around 2.9 V during the first cycle. According to previous studies the irreversible behavior was attributed to de-lithiation from the octahedral sites upon charging and migration of Mn^{3+} from the original octahedral sites to

Figure 2.8. SEM micrographs for (a) pristine o-LMO and (b) LMO/CNT composites. Reproduced from [117], copyright (2021) with permission from Elsevier.

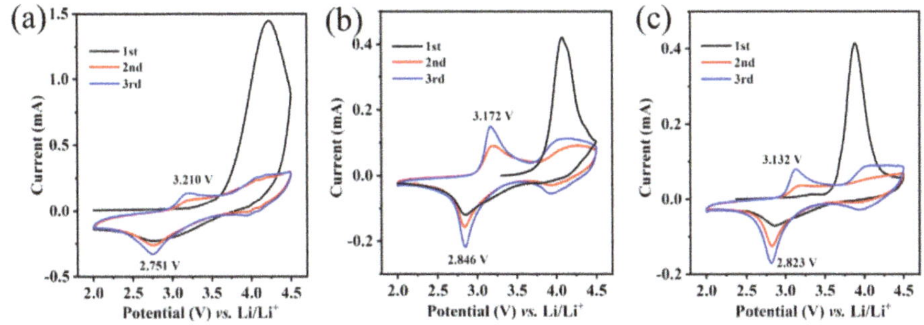

Figure 2.9. Cyclic voltammetry (CV) of pristine o-LMO and LMO/CNT composites. Reproduced from [117], copyright (2021) with permission from Elsevier.

neighboring vacant octahedral sites. As Li^+ cannot re-insert into its original octahedral sites, this results in formation of tetragonal $Li_2Mn_2O_4$. For further cycles, for pristine o-LMO, two less intense peaks at about 3.95 and 4.10 V were observed as a characteristic cubic spinel $LiMn_2O_4$ formation. However, no such peaks were observed for LMO/CNT composites. Thus, the addition of CNT in pristine o-LMO might have also played an important role for its structural stability.

The electrochemical characterization of pristine o-LMO and LMO/CNT composites was done in terms of charging–discharging curves, rate capabilities, and cycle performance. Figure 2.10((a) and (b)) represents the cycling performance and rate capability of the pristine o-LMO and LMO/CNT composites with 1% and 5% CNT. The LMO/CNT composites showed enhancement in the capacity and cycling stability significantly compared to pristine o-LMO, also it increased with CNT content from 1 to 5 wt%. A LMO/CNT composite with 5% CNT showed a high capacity of 204.9 mAh g^{-1} and capacity retention rate of 97.7% at 0.1 C after 50 cycles. Figure 2.10((c) and (d)) shows the charge–discharge profiles of pristine o-LMO and LMO/CNT composites with 5% CNT for the 1st, 5th, 10th, 20th, and 30th cycles. Both electrodes, pristine o-LMO and LMO/CNT composites, show a plateau at 3.6 and 3.0 V, which may be due to o-LMO. However, after 20 cycles, in the case of pristine o-LMO, the plateau at 3.0 V shortens, possibly as a result of the transformation of o-$LiMnO_2$ to spinel $LiMn_2O_4$ [118–121].

2.7.5 Nickel manganese cobalt (NMC) based positive electrode material

Even though LMO is an inexpensive and environmentally-friendly material, the specific capacity and cyclic performance is far from predicted values due to structural instability and sluggish lithium-ion diffusion kinetics. The structural instability basically comes from closer ionic radii of Mn^{3+} (in high spin configuration) to Li^+, which resulted in poor electrochemical activity. It has been observed that the addition of Ni and Co in the host lattice matrix of LMO showed high capacity with good retention over a large number of cycles. In view of this, modified LMO in the form of $LiNi_xMn_yCo_{1-x-y}O_2$ (NMC) has emerged as a promising candidate with high theoretical energy density 350 Wh kg^{-1} at cell level and a stable structure for future automotive and energy storage technology [122–125].

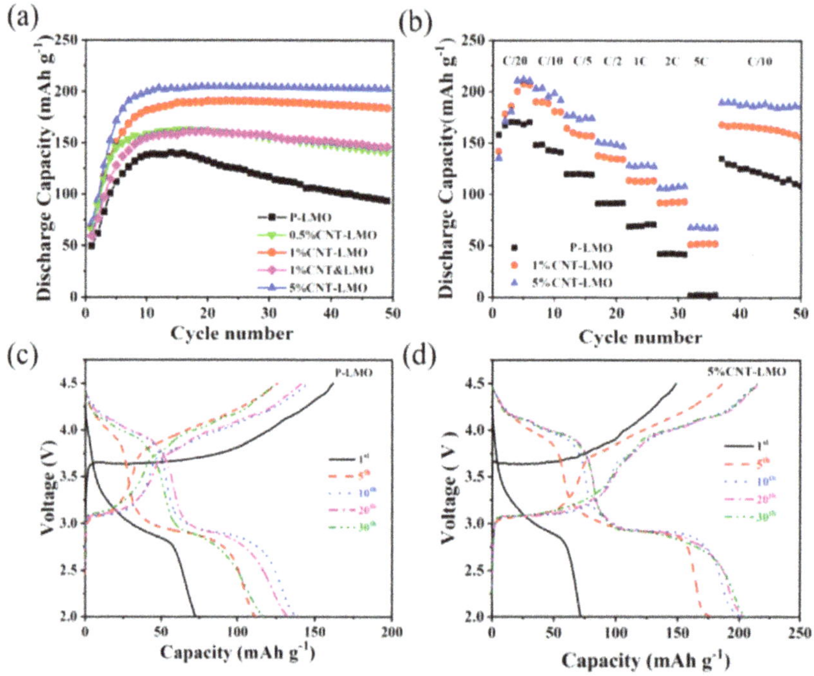

Figure 2.10. Electrochemical performance of LIBs with pristine o-LMO and LMO/CNT composites. Cycling performance and rate capability of: (a) pristine o-LMO and LMO/CNT composites with different % of CNTs, (b) pristine o-LMO and LMO/CNT composites with 1% and 5% CNT. Charge–discharge profiles of half cells for the 1st, 5th, 10th, 20th, and 30th cycles with (b) pristine o-LMO and (d) LMO/CNT composites with 5% CNT. Reproduced from [117], copyright (2021) with permission from Elsevier.

In NMC, nickel, manganese, and cobalt can be combined in different ways and in different proportions to increase electrochemical properties while suppressing the demerits of individuals. For instance, Ni is known for its high energy density, but leads to poor structural stability, Mn is beneficial for low internal resistance, good cyclic and thermal stability, but provides less specific energy, and the presence of Co reduces the degree of Li/Ni off-stoichiometry, but has a higher cost [16, 123–128]. The explored NMC compounds are shown in figure 2.11.

Early studies on NMC were focused on the Ni-rich composition ($0.5 \leqslant x \leqslant 1$), which showed effective reduction in off-stoichiometry; however, the energy density never achieved its theoretical limit 350 Wh kg^{-1} [125, 126]. For instance, Ohzuku *et al* prepared Li $Ni_{1/3}Mn_{1/3}Co_{1/3}O_2$ (NMC11) at 1000 °C in air, which showed a capacity of 200 mAh g^{-1} when charged to 5 V vs. Li^+/Li, and this is considered to be the first report on it [129]. Later, many investigations using various synthesis techniques were conducted; among them, synthesis by co-precipitation of mixed hydroxides, solid state, and sol–gel have all been thoroughly investigated from a commercial point of view. Different mixtures of hydroxides in co-precipitation were found to improve the cation distribution and phase purity of the material. The specific capacity for NMC111 in the range of 160 mAh g^{-1} to 200 mAh g^{-1} at a

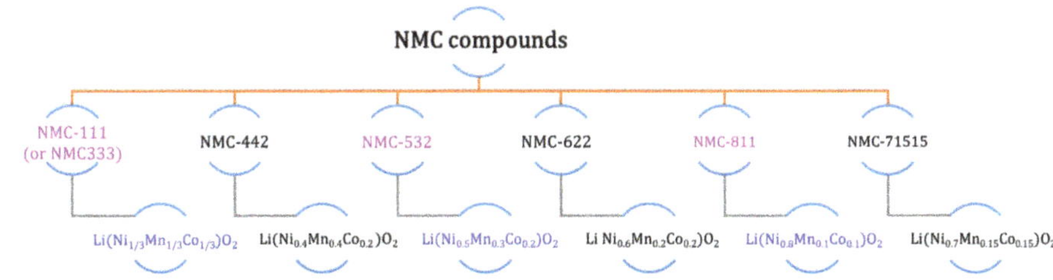

Figure 2.11. NMC cathode materials for LIBs.

cut-off voltage of 4.6 V vs. Li^+/Li was achieved. It has been reported that during the intercalation and de-intercalation process of NMC111, the redox process was mainly carried out by Ni^{2+}/Ni^{4+}, while Mn^{4+} looks electrochemically inactive and Co^{3+}/Co^{4+} does not significantly contribute to the redox process. This may result in surface reactions with oxygen release. According to compositions that have been reported, 4.3–4.4 V is considered to be the optimum range for the cut-off voltage for NMC111 [127–133]. Further, to reduce the material cost, NMC with composition $LiNi_xMn_xCo_{1-2x}O_2$ with $0.33 \leqslant x \leqslant 0.5$ has been examined [134, 135]. Among this class, $LiNi_{0.4}Mn_{0.4}Co_{0.2}O_2$ (referred to as NMC442) has been reported with a discharge capacity of 180 mAh g^{-1} and good electrochemical activities [136]. Ngala *et al* reported XPS analysis of NMC442, and showed that the oxidation states of Mn and Ni are mainly +4 and +2, respectively [134]. Like NMC111, here in NMC442, only Ni (Ni^{2+}/Ni^{4+}) and Co(Co^{3+}/Co^{4+}) significantly contribute to the redox process, while Mn^{4+} remains electrochemically inactive [137]. Thus, Ni-rich NMC was found to be attractive as it offers better electrochemical performance due to a significant contribution of Ni to the redox process as compared to Co and Mn. Nevertheless, the production of Ni-rich NMCs is complex and requires precision of the synthesis parameters. Several synthesis parameters such as annealing temperature, duration, and atmosphere were found to influence the Ni content, degree of cation mixing, surface texture and morphology, off-stoichiometry, and electrochemical active surface area [134, 138].

Recently, compositions with 40% Ni in NMC have also been studied to optimize the percentage of Ni. In such a case, a charge neutrality was observed as the effect of a Ni/Mn mismatch ratio due to partial oxidation of Ni^{2+} to Ni^{3+}. Among Ni-rich NMC, NMC532 is one of the most studied compositions, which exhibits a discharge capacity of about 214 mAh g^{-1} at 4.6 V and good retention rate. It has been observed that the Ni-rich composition $LiNi_xMn_yCo_{1-x-y}O_2$ with $x = 0.6$–0.8 showed enhancement in the discharge capacity, but in a moderate potential window. It also leads to an off-stoichiometry problem and thermal instability of NMC [139–143]. Furthermore, several groups from both industry and academia argued that next-generation cathode materials with a specific capacity of more than 200 mAh g^{-1} can be achieved through NMC with composition like $LiNi_{0.6}Mn_{0.4-x}Co_xO_2$. An increase in Ni may help to enhance the Ni^{3+}/Ni^{2+} ratio and hence increase the specific capacity [144].

Among these compositions, $LiNi_{0.6}Mn_{0.2}Co_{0.2}O_2$ (NMC622) has been explored initially, which delivers a discharge capacity of 172 mAh g^{-1} at 4.3 V. The equal content of Co and Mn results in minimization of polarization and enhancement in capacity retention of about 94% after 100 cycles [145–147].

Nevertheless, a further increase in Ni in NMC with $x = 0.7$ and $x = 0.8$ showed discharge capacities in the range of 180 to 200 mAh g^{-1} at a cut-off voltage of 4.3 V vs. Li^+/Li. However, along with the electrochemical benefits, more Ni in NMC also showed some problems during Li extraction/insertion like thermal instability, cation mixing, and parasitic reactions. This resulted in pressurized internal environments and continuous degradation of cell components [145–148]. The Li extraction/insertion reaction mitigated the cation mixing of the highly reactive de-lithiated cathode surface at high potentials. Currently, a $LiNi_{0.8}Mn_{0.1}Co_{0.1}O_2$ (NMC811, with x=0.8) composition with a maximum of about 80% Ni, 10% Mn, and 10% Co exhibited a discharge capacity of about 200 mAh g^{-1} at 4.5 V vs. Li^+/Li. However, it also possesses some unavoidable issues of thermal instability and low cyclic performance [142, 149–151].

2.7.6 Spinel $LiMn_2O_4$-based positive electrode material

Another major class of cathode material is the spinel structure material $LiMn_2O_4$ (s-LMO), which is considered to be a promising alternative as a positive electrode material for LIBs due to economic and environmental reasons. It has low cost, good thermal stability, good electronic as well as chemical properties, non-toxic nature, excellent safety, and three-dimensional Li^+ diffusion pathways [29, 152]. The spinel $LiMn_2O_4$ belongs to space group *Fd*-3*m* (cubic symmetry) with repeatable distribution of cations within the anionic network and has substantial stability during the process of lithium-ion insertion/extraction. In s-LMO, Li occupies the tetrahedral 8a sites, Mn occupies the octahedral 16d sites, and oxygen the 32e sites, as shown in figure 2.3(b) [38]. This forms a cube-like structure with alternate 3D manganese-oxide (Mn_2O_4) framework which allows insertion of the Li ion. It has been reported that the disordered spinel phase (space group $Fd_{-3}\,m$) is more stable and hence showed a much better rate capability than that of the ordered one (space group $P\,4_332$) [30, 153]. Figure 2.5(b) illustrates the ion diffusion path in spinel $LiMn_2O_4$. During lithium-ion insertion, the Li ions occupy the interstitial spaces defined by the 3D manganese-oxide (Mn_2O_4) polyhedral framework and diffuses amongst two succeeding octahedral sites through a tetrahedral site with a diffusion rate of about 10^{-6}–10^{-10} $cm^2\ s^{-1}$ [38]. Initially, when lithium is extracted from the 8a tetrahedral sites at around the 4.0 V region, the structure remains intact due to its cubic spinel symmetry. However, the transition from cubic to tetragonal occurs during lithium extraction/insertion from 16c octahedral sites at around the 3.0 V region due to a change in the oxidation state of Mn. This transition led to a rise in the lattice parameter ‘c/a’ ratio by 16%, and volume of the unit cell by 6.5%; which disrupts the structural symmetry. The tetragonal phase is more disordered and low symmetry phase. Thus, limited better performance can be achieved with s-LMO in a region around 4 V, while structural transition leads to rapid capacity fade at around 3.0 V [154–156].

Several other factors along with structural transition were also found to be responsible for capacity fading [155–158]. These are:

- Instability in the electrolyte at high temperature (>55 °C) and at higher operating voltage (>4.3 V);
- Dissolution of Mn in the electrolyte due to disproportionate translation of Mn^{3+} into either Mn^{2+} or Mn^{4+};
- Subsequent deposition of Mn on the anode material surface during dissolution;
- Jahn–Teller distortion due to Mn^{3+} at the octahedral sites;
- Electrode–electrolyte interface reactions;
- Particle size and its distribution, surface morphology, and effective specific surface area.

Numerous methods such as hydrothermal, co-precipitation, sol–gel, solid state, spray drying, molten salt, oxidation ion-exchange method, etc, have been well reported for the synthesis of s-LMO. Several approaches have tried to overcome the capacity fading problem due to structural transition and to improve its high voltage and high temperature performance. Among several efforts, metal oxide coating, doping with metal ions, fine particle size, and surface morphologies were found to be effective for improvement in electrochemical performance. For instance, doping with selected metal ions (B, Zn, Ti, Al, Co, Cr, Fe, Y, Nb, Mo, Ru, Mg, Ni, Mg, Zn, Ga, Nd, Sm, Gd, Tb, etc) in s-LMO resulted in a decrease in Jahn–Teller instability by preventing Mn^{2+} dissolution [159–166].

Surface modification with metal oxides such as Al_2O_3, NiO, Co_3O_4, MgO, ZnO, $Li_4Ti_5O_{12}$, K_2CO_3, Na_2CO_3, and V_2O_5 suppresses the dissolution of Mn in the electrolyte [44, 167, 168]. Surface coating with carbonates such as K_2CO_3, Na_2CO_3, and Li_2CO_3 causes the neutralization of acid in the electrolytes, and hence can help to prevent the dissolution of Mn into the electrolyte via the coating layer and degradation of the electrolytes effectively. $Li_4Ti_5O_{12}$-coated $LiMn_2O_4$ showed improved electrochemical performance along with more cycle life. The enhanced performance was attributed to high thermal stability and higher chemical diffusion coefficient (10^{-6} cm^2 s^{-1}) of $Li_4Ti_5O_{12}$ than $LiMn_2O_4$ (10^{-10}–10^{-12} cm^2 s^{-1}) [168]. The ordered mesoporous materials and 3D surface morphology (nanotubes, nanorods, nanowires, microspheres, etc) also showed an enhanced electrochemical performance with a discharge capacity of about 117.2 mAh g^{-1} at 1 A g^{-1} due to ease of diffusion of Li ions and relatively large surface area with more active electrochemical sites [169–171]. Composites of s-LMO with carbon nanotubes and graphene also showed improved performance due to enhancement in electrical conductivity and channelized charge transport rate [172]. Recently, studies on a nanocomposite of s-LMO with NMC showed much better electrochemical performance as compared to their individual form with prolong cycle life, making it suitable for future EV application. The enhanced performance was attributed to a synergistic effect of their properties, resulted in high capacity and better capacity retention [173]. The electrochemical performance of LCO, LNO, NCA, o-LMO, NMC, s-LMO, their different nanostructures and composites are tabulated in table 2.2 [114, 117, 174–191].

Table 2.2. Electrochemical performance of LCO, LNO, NCA, o-LMO, NMC, s-LMO; their different nanostructures and nanocomposites.

Cathode material	Specific discharge capacity mAh g^{-1} at // mAg^{-1}@C	% of capacity retention at // mAg^{-1}@C	Cycles	References
Ti-doped LCO	205	97 @ 0.1	200	[67]
LCO (Solid-state method)	98.3	80.7 @ 1.0	50	[72]
LCO (Template-assisted sol–gel)	166	81		[73]
Pristine LCO	127	72		[73]
$LiCo_XMn_{2-X}O_4$ @ LCO	143.8	66.1	300	[77]
LCO	209.3	89.3	100	[93]
$LiNiO_2$	194.5 @ 2.0	85.5 @ 2.0	300	[174]
$LiNiO_2$	231.7 @ 0.1	91.3 @ 0.1	400	[175]
Zr-doped $LiNiO_2$	246.5 @ 0.1	81 @ 0.1	100	[176]
Co-free $LiNiO_2$	211.9 @ 0.1	65.84 @ 0.5	50	[177]
Tungsten-doped $LiNiO_2$	195.6 @ 0.5	73.7 @ 0.5	100	[178]
$LiNi_{0.88}Co_{0.09}Al_{0.03}O_2$ (NCA)	154.6 @ 0.1	75.93 @ 0.1	40	[179]
Y_2O_3@NCA	172 @ 0.5	88 @ 0.5	100	[180]
G@Y_2O_3@NCA	180 @ 0.5	92 @ 0.5	100	[180]
Ni-rich NMC	197 @ 10	93 @ 10	100	[181]
Cr_2O_3@$LiNi_{0.8}Co_{0.15}Al_{0.05}O_2$	162.2 @ 0.5	92.1 @ 0.5	50	[182]
Mesoporous $LiMnO_2$	191.5 @ 0.1	84.9 @ 0.1	50	[114]
Li_2MnO_3–$LiMnO_2$	265 @ 0.1	93 @ 0.5	80	[183]
$LiMnO_2$@CNTs	204.9 @ 0.1	97.7 @ 0.1	50	[117]
$LiMnO_2$@rGO	185.6 // 100	80 // 100	100	[184]
Graphite@NMC 111 @MWCNTs	150.7 @ 1.0	30.4 @ 1.0	1000	[185]
NMC111@MWCNTs	118.5 @ 1.0	44.2 @ 1.0	1000	[185]
NMC111	84.9 @ 1.0	59.2 @ 1.0	1000	[185]
NMC622	153.60 @ 0.5	70.9 @ 0.5	100	[186]
Lithium hydroxide-NMC622	177 @ 0.05	88 @ 0.5	30	[187]
Lithium acetate-NMC622	181 @ 0.05	91 @ 0.5	30	
Lithium nitrate-NMC622	180 @ 0.05	93 @ 0.5	30	
Lithium carbonate-NMC622	187 @ 0.05	91 @ 0.5	30	
Mg^{2+} and Ti^{4+} co-doped spinel $LiMn_2O_4$	124 @ 0.1	97 @ 5.0	100	[188]
La–Sr–Mn–O@$LiMn_2O_4$	129.9 @ 0.1	90.6 @ 1.0	500	[189]
$LiMn_2O_4$@C	83 // 2	92 // 2	200	[190]
Al–F co-doped spinel $LiMn_2O_4$	115.5 @ 0.1	80 @ 0.1	~350	[191]

2.7.7 Olivine $LiFePO_4$-based positive electrode material

Among the ordered olivine phosphate material family, $LiFePO_4$ has been extensively investigated as the third best promising compound for cathode material after LCO and LMO. It is of particular interest because of its good thermal stability, abundancy, low cost, non-toxicity, resistance to overcharge, environment-friendly nature, no memory effect, good safety, and stable voltage plateau [32]. It is also known as polyanionic compounds composed of a corner-shared octahedron (FeO_6) with central iron surrounded by six oxygen atoms, edge-shared tetrahedron (PO_4) with central phosphorus surrounded by four oxygen atoms, and edge-shared octahedron (LiO_6) with central lithium surrounded by six oxygen atoms, as shown in figure 2.3(c). The octahedron (FeO_6) shares edges with two LiO_6 octahedra and one PO_4 tetrahedron, resulting in a three-dimensional zigzag structure. A schematic of diffusion routes in olivine-structured phosphates $LiMPO_4$ as calculated by DFT methods is shown in figure 2.5(f) [20, 39, 192]. The approximate tetrahedral site between the two octahedral sites in the olivine structure serves as the transition state for lithium diffusion along the chain. According to Islam *et al*, a nonlinear, curved pathway connecting Li sites facilitates Li-ion mobility in the one-dimensional [010] channel of the olivine crystal structures [193]. Later, this mechanism was empirically validated by Nishimura *et al* [194].

Furthermore, it possesses a theoretical specific capacity of 170 mAh g^{-1}, working potential of 3.4 V, theoretical density of 3.6 gcm^{-3}, good rate capability, and low volumetric energy density of about 220 Wh L^{-1}, making it a suitable alternative for layered NMC, NCA for next-generation EVs [195]. However, owing to a low intrinsic electronic conductivity of about 10^{-9} S cm^{-1} and low diffusion coefficient of lithium ions of about 10^{-14} cm^{-2} s^{-1} this leads to a remarkable self-discharge and poor rate capability [196]. Yang *et al* investigated the diffusion process of Li ions in LFP theoretically using first principle-based calculations. According to the calculations reported by them, the types of defects, Li-ion concentration, lattice strain, particle size, and distribution in the material have great influence on the Li-ion diffusion coefficient [197].

Numerous studies have been carried out to enhance the electrical and ionic conductivity to obtain better cyclic performance and rate capability [198–200]. Among the several approaches, reduction of particle size, doping of metal cations, and coating or composites of LFP with conducting carbon material showed improved performance [201, 202]. Doping of metal cations such as Zr, Nb, or Mg in LFP exhibited improved electrochemical performance, rate capability, and power density due to higher electrical conductivity and a reduction in defect level.

It has been reported that nanoparticles (<50 nm) with narrow distribution showed excellent rate performance due to suppressed defects in the material. Several synthesis routes including co-precipitation, solid-state reaction, hydrothermal, mechanical activation, microwave, and sol–gel have been adopted for development of LFP with a nanoporous structure [203–207]. Further, the carbon-coated LFP nanoparticles with few wt% showed much better performance as the polarization of the electrode is inhibited and also the electrical conductivity is enhanced [20, 208]. Despite these efforts, the energy density of LIBs with a LFP cathode is still in the range of about ~120 Wh kg^{-1}, due to limited ionic and electrical conductivity [109, 209].

2.7.8 Other phosphates with an olivine structure as the cathode for LIBs

Other phosphates with an olivine structure including $LiMnPO_4$, $LiNiPO_4$, and $LiCoPO_4$ have been explored for cathodes in LIBs [210–212]. $LiCoPO_4$ has a higher working potential of 4.8 V vs. Li^+/Li than $LiFePO_4$ and theoretical capacity of 167 mAh g^{-1} [213]. However, it exhibits a low lithium-ion diffusion rate and poor cyclic performance due to low electronic conductivity ($<10^{-9}$ S cm^{-1}). Also, the higher working potential than the potential window for the commercial electrolyte causes its decomposition [214–216]. Similarly, $LiNiPO_4$ also has a high operating potential of about 5.1 V vs. Li^+/Li potential window for the commercial electrolyte, which limits its application [212].

However, $LiMnPO_4$ has an operating potential of 4.1 V vs. Li^+/Li within the stability window of the commercial electrolytes. It is considered as the second best material from the olivine family as it has more thermal stability, is non-toxic, low cost, environmentally friendly, with a higher working potential and high theoretical capacity of 171 mAh g^{-1} [217]. Regardless of these properties, it showed relatively poor cycling stability and low lithium-ion diffusion rate due to low electronic conductivity ($<10^{-10}$ S cm^{-1}) and Jahn–Teller distortion and interface volume strain during the de-intercalation/intercalation process [218–223].

2.7.9 Iron and vanadium based fluorophosphates as cathode materials

According to recent studies, novel cathode materials for LIBs based on the widely used structural compounds fluorophosphates and fluorosulphates, are based on iron and vanadium [20, 224–227]. The major compounds from this class include lithium iron sulfate fluoride ($LiFeSO_4F$) [224–226], lithium vanadium fluorophosphate ($LiVPO_4F$), lithium vanadium phosphate ($LiVPO_4$) [227, 228], and lithium vanadium phosphate oxide ($LiVPO_4O$) [229–231], which have been well studied as promising cathode materials due to their structural stability, high working potential, and supporting structure for simultaneous diffusion of two Li ions during the insertion/extraction process. For instance, ideally a very high specific capacity of 312 mAh g^{-1} can be achieved with $LiVPO_4F$ by exploiting more than one oxidation state from +2 to +5 of vanadium. But in practice the performance is poor, since it is not possible to utilize more than two redox couples simultaneously, as each has a different redox potential [227]. $LiFeSO_4F$ has a high working voltage of about 3.9 V vs. Li^+/Li and more electronic and ionic conductivity, resulting in high energy densities comparable with LFP and $LiVPO_4$ [20, 232].

Table 2.3 shows the comparative performance of all explored phosphates and their composite cathode materials [219, 233–250].

Table 2.3. Comparative performance of all explored phosphates and their nanocomposite cathode materials.

Cathode material	Specific discharge capacity mAh g^{-1} at // mAg^{-1}@C	% of capacity retention at // mAg^{-1}@C	Cycles	References
$LiFePO_4$	594@2.0	83.9@0.2	400	[233]
$LiFePO_4$@G	117.4 // 25	78 // 100	1000	[234]
$LiFePO_4$@C	165.3@0.2	98@0.2	100	[235]
Zn-doped $LiFePO_4$	120.0@0.25	60.9 @1.0	200	[236]
$LiFePO_4$@C	159.8@2.0	84.2@2.0	500	[237]
Cellulose nanofiber derived C@rGO@$LiFePO_4$	168.9@0.1	91.5@10	500	[238]
$LiMnPO_4$	130@1.0	100@1.0	200	[239]
Ti-doped $LiMnPO_4$@$NaTi_2(PO_4)_3$ @C@3D G	164.8@0.05	93.3@2.0	600	[240]
V-doped $LiMnPO_4$	126@0.2	60@5.0	50	[241]
Er-doped $LiMnPO_4$@C	142.5@0.05	90.8@1.0	100	[242]
$LiMnPO_4$@C	161.5@0.05	96.4@10	200	[243]
$LiMnPO_4$ nanorods@C	156.8@0.05	98.2@0.2	100	[244]
$LiAlO_2$@$LiMnPO_4$	105@10	98.4@10	100	[219]
$LiMnPO_4$@$Li_3V_2(PO_4)_3$@C@rGO	116.3@10	85.2@10	500	[245]
$LiNiPO_4$@C	175@0.05	82.7@0.05	100	[246]
$LiCoPO_4$	135@0.05			[247]
$LiCoPO_4$@$MnCo_2O_4$	95@0.1	74@0.1	50	[248]
$LiCoPO_4$@AlF_3	159@0.1	91@0.1	50	[249]
$LiCoPO_4$@C nanoflake	154.6@0.1	93.1@0.1	100	[250]

2.7.10 Polyoxyanion compounds (silicates): Li_2MSiO_4-based cathode materials

Silicate-based polyoxyanion compounds have emerged as a new class of cathode material for high voltage LIBs. The general formula is Li_2MSiO_4, where M is the metal cation, and can be Fe, Mn, Co, and Ni. Just like polyanionic material, the cations occupy the tetrahedral and octahedral sites and undergo transition into different phases during the extraction/insertion process as shown in figure 2.3(d). With high working potential (>4.5 V) it supports simultaneous diffusion of two Li ions during the insertion/extraction process resulting in high theoretical capacity of above 330 mAh g^{-1} [20, 36, 251]. All four silicates, Li_2FeSiO_4, Li_2MnSiO_4, Li_2CoSiO_4, and Li_2NiSiO_4 were extensively studied for the application of energy storage devices.

Among these, the lithium de-intercalation/extraction voltages for the first and second lithium ion of Li_2CoSiO_4 were found to be in the range of 4.5 to 5.0 V, whereas for Li_2NiSiO_4 they were in the range of 4.67 to 5.12 V. Because of the high voltage (>4.5 V), which is not in a stability potential window of commercially available organic electrolytes, Li_2CoSiO_4 and Li_2NiSiO_4 are not suitable for practical applications. High operating potential of the cathode material causes decomposition of the electrolyte [36, 252].

Li_2FeSiO_4 has attracted the attention of researchers as a cathode material for LIBs as it is composed of metal cations such as silicon and iron, which are non-toxic, abundant, thermally stable, environmentally friendly, and also provide high theoretical reversible capacity with good cyclic performance [20, 35, 251, 253].

By varying synthesis conditions, Li_2FeSiO_4 can be synthesized in different phases that belong to three different space groups: orthorhombic ($Pmn2_1$), inverse-orthorhombic ($Pmn2_1$), orthorhombic (*Pmnb*), and monoclinic ($P2_1$/n). All cations Li^+, Fe^{2+}, and Si^{4+} occupy the tetrahedral sites, forming tetrahedra polymorphs LiO_4, FeO_4, and SiO_4, respectively, with oxygen atoms shown in figure 2.12 [37, 251, 254–257].

The arrangement of cations/anions, filling of the tetrahedra and their covalent nature decides the structural, chemical stability, and Li-ion diffusion rate during the lithiation/de-lithiation processes. The first oxidation potential, i.e. lithium de-intercalation/extraction voltage, for the first lithium ion of orthorhombic *Pmnb*—Li_2FeSiO_4 and $Pmn2_1$—Li_2FeSiO_4 and $P2_1/n$-based Li_2FeSiO_4 is 3.06, 3.13, and 3.10 V vs. Li^+/Li, respectively, making them suitable for practical applications [255–257]. Among all of the explored structural phases of Li_2FeSiO_4, the $Pmn2_1$ orthorhombic phase showed good cyclic behavior with optimum electrochemical performance due to ease of Li-ion diffusion channels and comparatively good thermal stability. However, the electrochemical performance is also decided by the high working potential responsible for simultaneous Fe^{2+}/Fe^{3+} and Fe^{3+}/Fe^{4+} redox couples during the charge–discharge process [252].

However, the inhibited second Li-ion extraction due to the high oxidation potential of about 4.5 V, low electronic conductivity and Li-ion diffusion rate of about 10^{-14} cm^2 s^{-1}, Li_2FeSiO_4 with all its phases showed poor cyclic behavior and energy density. Numerous studies have been carried out to overcome these limitations and to obtain better cyclic performance with high energy density. Among several approaches, reduction in particle size, doping with metal cations, and coating Li_2FeSiO_4 surface with conducting carbon-based materials were found to be effective. Reduction in particle size causes a decrease in internal resistance and

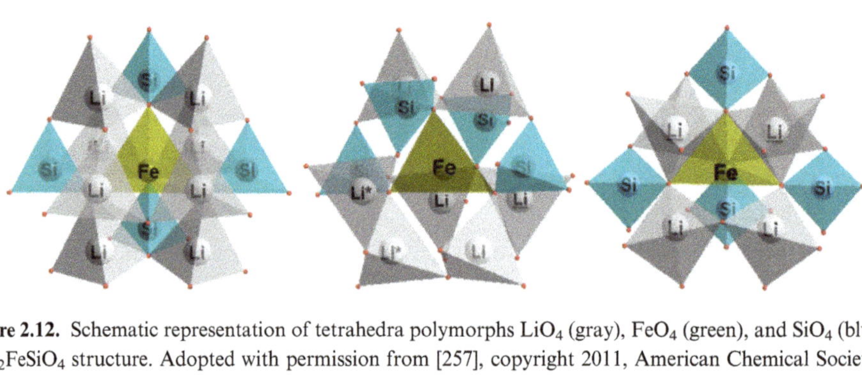

Figure 2.12. Schematic representation of tetrahedra polymorphs LiO_4 (gray), FeO_4 (green), and SiO_4 (blue) in a Li_2FeSiO_4 structure. Adopted with permission from [257], copyright 2011, American Chemical Society.

Li-ion diffusion path length, while carbon coating enhances performance due to an increase in effective surface area and electrical conductivity [35, 36, 37, 251, 253].

For example, Deng *et al* prepared a $Li_2FeSiO_4xLi_3PO_4/C$ composite by a citric acid-assisted sol–gel method. The calcination temperature greatly influenced the electrochemical performance. The authors reported that the low calcination temperature revealed better energy density, which was attributed to higher carbon content, smaller particle size, and uniform distribution of pores. Doping of Ni in Li_2FeSiO_4 occupying the tetrahedral site of Fe showed enhanced electrochemical performance with a discharge capacity of 160.1 mAh g^{-1}. The enhanced performance was attributed to an increase in Li-ion diffusion coefficient from 1.38×10^{-12} to 3.34×10^{-12} cm^2 s^{-1} after doping with Ni [258].

A composite electrode with amorphous carbon, MWCNTs, and nanosized Li_2FeSiO_4 showed a much higher discharge capacity of 310 mAh g^{-1} at 0.1 C. The sample also showed good rate capability with a capacity of about 50 mAh g^{-1} even at high rate 50 C after 1000 cycles. The better electrochemical performance with a high discharge capacity was attributed to the synergistic impact of higher Li-ion diffusion kinetics and electrical conductive networks created by graphene sheets [259].

Li_2MnSiO_4 is the second most explored silicate after Li_2FeSiO_4 and comparatively the most preferred due to multiple oxidation states of Mn, and offers two redox couples (Mn^{2+}/Mn^{3+} and Mn^{3+}/Mn^{4+}). Moreover, it has oxidation potentials for extraction of first and second lithium of about 4.2 and 4.4 V, respectively, which lies in the stability window of the electrolyte. It also has crystal structures with a monoclinic phase belonging to $P2_1/n$ and Pn space groups and the orthorhombic phase belongs to $Pmn2_1$ and $Pmnb$ space groups [260].

All these four polymorphs have cations Li^+, Mn^{2+}, and Si^{4+} occupying tetrahedral sites in a distorted hexagonally close-packed oxygen array. The difference in crystal structure of Li_2FeSiO_4 and Li_2MnSiO_4 lies in the arrangement of tetrahedra LiO_4, MnO_4, and SiO_4. The channelized three-dimensional framework with Li ions, high operating voltage, and multiple oxidation of Mn results in a discharge capacity of more than 300 mAh g^{-1}. However, intrinsic low electrical conductivity ($<10^{-16}$ S cm^{-1}) and phase variation during the charge–discharge process limits its practical application. It has been reported that the Jahn–Teller structural instability due to decomposition of Mn^{3+} ions is mainly responsible for the phase variation [37, 161]. Various approaches have been developed to overcome these obstacles, among which doping with metal cations, reduction in particle size, surface morphology, and conducting organic polymers or carbon-based material coating were found to be helpful for improvement in the electrochemical performance [262–264].

Li_2MnSiO_4 can be synthesized by various physical and chemical methods, among which sol–gel is reported as a versatile choice for the preparation of nanosized Li_2MnSiO_4 particles. The synthesis of Li_2MnSiO_4 nanoparticles by a solvothermal method at 300 °C results in a smaller particle size of about 15 nm and a higher specific surface area. A Li_2MnSiO_4/C/graphene composite electrode revealed a discharge capacity of 215.3 mAh g^{-1} at 0.05 C and good retention with reversible capacity of 175 mAh g^{-1} after 40 cycles. A composite electrode with conductive polymer (poly(3,4-ethylenedioxythiophene) [PEDOT])-coated Li_2MnSiO_4 showed a high discharge capacity of about 313 mAh g^{-1} attributed to the high electrical conductivity offered by the PEDOT and uniform distribution of fine nanoparticles [265].

A composite electrode with carbon and CNT-embedded Li_2MnSiO_4 showed a discharge capacity of about 200 mAh g^{-1} attributed to the high specific surface area, moderate thermal and electrochemical stability, effective Li^+ ion diffusion rate, and high electrical conductivity [266]. Furthermore, doping of Li_2MnSiO_4 with metal cations such as Al, Cr, and Ga occupying the tetrahedral sites in the structure was found to be favorable for good cyclic stability and excellent electrochemical performance [265, 267]. The performance of silicate-based cathode materials is shown in table 2.4 [268–283].

2.7.11 Other positive electrode materials

Layered lithium vanadium oxide in different configurations like LiV_2O_5, LiV_3O_8, and LiV_6O_{13} has been explored as a cathode material for LIBs [284, 285]. Among these, LiV_2O_5 has a higher energy density of about ~450 Wh kg^{-1} and is expected to give a better cyclic performance. However, low intrinsic electric conductivity, low operating potential, significant vanadium dissolution to V^{4+} and V^{3+}, and slow Li-ion diffusion rate limit further exploration. LiV_3O_8 has a good specific capacity of about 280 mAh g^{-1}, but has a low diffusion coefficient of about 10^{-13} cm^2 s^{-1} [286]. Research has been focused on improving structural stability, Li-ion diffusion kinetics, electrical conductivity, and excellent cycling performance. Several cathode modification techniques were used to achieve the increased performance including, doping with

Table 2.4. Electrochemical performance of silicates and their nanocomposite cathode materials.

Cathode material	Specific discharge capacity mAh g^{-1} at // mAg^{-1}@C	% of capacity Retention (cycle) at // mAg^{-1}@C	Cycle	References
Li_2FeSiO_4	175.9@0.06	80@0.06	65	[268]
Li_2FeSiO_4@C	174@0.1	95.3@0.1	100	[269]
Li_2FeSiO_4@C@Ag	152@0.2	97.9@1.0	100	[270]
Li_2FeSiO_4@C	243.5	95.2	200	[271]
Mg-doped Li_2FeSiO_4	142.3@0.1	96@0.1	100	[272]
Li_2MnSiO_4	250@0.1	48.4@0.1	200	[273]
Li_2MnSiO_4@C	230 // 50	81 // 50	100	[274]
Cr-doped Li_2MnSiO_4@C	255		50	[275]
Li_2MnSiO_4@NC	277.3@0.1	74.4@0.1	50	[276]
C@mesoporous-Li_2MnSiO_4	217			[277]
Li_2CoSiO_4	135			[278]
Li_2CoSiO_4@C	101@0.05	90@0.05	10	[279]
Li_2CoSiO_4@C	136.6@0.1	91@0.1	50	[280]
V-doped Li_2CoSiO_4	220.1			[281]
Li_2CoSiO_4@C	135.0			[281]
Al-doped Li_2CoSiO_4@C	140			[282]
P-doped Li_2CoSiO_4@C	144			[283]

metal cations such as Ti, Cr and Mo, composite with graphene and other carbon materials, and surface coating by conducting polymers like polypyrrole and polyaniline. [287, 290–293]. In spite of these attempts to improve the performance, they were unable to meet the remarkable improvement in the Li-ion diffusion rate and structural stability at high operating voltage.

Among transition metal chalcogenides, MoS_2, MnS_2, VS_2, FeS_2, WS_2, NbS_2 and TiS_2 have also been used as cathode materials for LIBs [23–25, 294, 295]. The advantage with these materials lies in their high electrical conductivity, abundancy, low cost, and easy synthesis routes. However, the lower potential window, structural instability, and poor Li-ion diffusion kinetics hinder their practical applications [296–298].

Like silicates, newer classes of material with polyoxyanion groups $(BO_3)^{3-}$ have emerged for cathode materials in LIBs, known as borates. They have a general formula $LiMBO_3$, where M is the metal cation and can be any one of Fe, Co, Ni, and Mn [299]. All four borates $LiFeBO_3$, $LiMnBO_3$, $LiCoBO_3$, and $LiNiBO_3$ showed a theoretical capacity near to about 220 mAh g^{-1}. Among these, $LiFeBO_3$ and $LiMnBO_3$ were extensively studied for the application of energy storage devices. The oxidation potential for the first and second Li ion of $LiFeBO_3$ was found to be 3.1 and 2.9 V vs. Li^+/Li, which is in the stability potential window of the commercially available organic electrolytes. Its electrical conductivity is also higher than $LiFePO_4$ and has a value of about 3.9×10^{-7} S cm^{-1}. In spite of these superior properties, the electrochemical performance of $LiFeBO_3$ was still not able to reach the theoretical limit because of internal resistance, a lower energy density caused by its low operating potential, loss of Li, and degradation of active material due to partial oxidation of Fe resulting from air or moisture exposure [300, 301]. Another borate, $LiMnBO_3$, has attracted the attention of researchers as a cathode material for LIBs as it is non-toxic, abundant, thermally stable, and also exists in two phases: monoclinic $LiMnBO_3$ and hexagonal $LiMnBO_3$ [302]. However, a wide operating potential window of 1 to 4.8 V vs. Li^+/Li causes degradation of the electrolyte due to the conversion reaction, resulting in poor cyclic performance [303].

Similarly, if Si in silicate is replaced by Ti, this results in a new series of compounds known as titanates, having a general formula Li_2MTiO_4 (M = Mn, Fe, Co, and Ni). Among this series of titanates, two compounds, Li_2FeTiO_4 and Li_2MnTiO_4, have been studied as cathode materials for LIBs. Because of the strong bonding of Ti with O, and theoretical capacity of about 148 mAh g^{-1}, good rate capability and cycling performance can be expected. However, due to poor ionic and electronic conductivities they showed a discharge capacity near to 123 mAh g^{-1} [304]. The electrochemical properties of cathode materials including layered lithium vanadium oxide, metal chalcogenides, borates, and titanates are presented in table 2.5 [23, 24, 288, 304–334].

2.8 Summary

The discussion in this chapter provides insight into cathode material selection criteria, active cathode materials, interphases in lithium-ion batteries, and

Table 2.5. Electrochemical performance of layered lithium vanadium oxide, metal chalcogenides, borates, and titanates.

Cathode material	Specific discharge capacity mAh g^{-1} at // mAg^{-1}@C	% of capacity retention at // mAg^{-1}@C	Cycles	References
LiV_2O_5 nanorod	250.1 // 50	84.2 // 300	50	[305]
γ-LiV_2O_5	269.3 // 30	84.7 // 30	20	[306]
Nb^{5+}-doped LiV_3O_8	401@0.1	99.7@0.1	500	[307]
Cr-doped LiV_3O_8	254.8 // 150	94.4 // 150	100	[288]
1T@2H $MoSe_2$	843 // 100	99 // 100	100	[308]
WSe_2	982 // 100	83.28 // 100	100	[309]
1T-MoS_2@C	920.6 // 1000	95 // 1000	455	[310]
WS_2	589@0.1	60.1@0.1	60	[311]
ZnS@Graphite	117 // 3000	80 // 3000	50	[312]
VS_2@G	180.1@0.2	89.3@0.2	200	[313]
VS_2 sheets	180 // 200	100 // 200	400	[23]
TiO_2-B@VS_2	365.4@1.0	129.9@1.0	500	[314]
α-MnS@Co_9S_8@C	608 // 500	73 // 500	300	[315]
FeS_2@S-doped C	442.5 // 5000	89.14@1000	300	[316]
TiS_2–MWCNT	450 // 100	80 // 100	50	[317]
C@$LiFeBO_3$	367@0.05	85@0.05	50	[318]
G enhanced $LiFeBO_3$@C	189.6@0.1	95@1.0	50	[319]
Ni-doped $LiFeBO_3$@C	201.5@0.05	90.1@0.05	105	[320]
PVP@$LiMnBO_3$	162@0.05	60@0.05	20	[321]
$LiMnBO_3$ nanobeads	67@0.6	88@0.6	100	[322]
Li_2MnTiO_{4+z}	200@0.2	90@0.2	60	[323]
$LiMnBO_3$@C	148.2@0.05		50	[324]
$LiMnBO_3$	678.3	80.9	20	[325]
$LiCoBO_3$@C	76.7@0.1	65.9@0.1	50	[326]
$LiCoBO_3$@Ketjen Black	40@0.05		25	[327]
$LiCoBO_3$	160 // 98		52	[328]
$LiMnBO_3$–$LiCoBO_3$	60@0.05			[329]
Li_2FeTiO_4	367.8@0.1		20	[330]
Li_2FeTiO_4	153.8@0.1			[331]
Li_2FeTiO_4	123@0.05			[304]
Li_2FeTiO_4@C	226.1 // 2000		1500	[332]
Li_2FeTiO_4@G	218.6@1.0		200	[333]
Li_2MnTiO_4	187.5 // 30			[334]

component roles in LIBs. Along with reviewing materials for cathodes with an emphasis on their present trends and perspectives, it also underlines the essential requirements of each LIB component.

Less expensive insertion materials with higher energy and power densities and greater thermal stability have been discovered, especially for positive electrodes. The preferred class of materials for cathodes in LIBs are layered lithium metal oxides; but spinel, olivine, silicates, transition metal chalcogenides, borates, and titanates have also demonstrated their viability through effective incorporation of significant amounts of lithium and structural reversibility during intercalation and de-intercalation processes. Some of the kinetic processes that influence the capacity and rate of lithiation and de-lithiation in LIBs include lithium-ion transport and movement in the electrode materials, ion transport in the non-aqueous electrolyte, electron transfer at the electrode and electrolyte interface, and phase transition in the electrode materials brought on by lithiation or de-lithiation.

The modification of cathode materials, either through heteroatom doping or nanocomposites with carbon derivatives, was found to improve both the kinetics of lithium-ion diffusion and overall electrical conductivity. Improvement in structural stability was observed as a result of the cation doping, decreasing the Jahn–Teller effect. Mesopores considerably enhance the specific surface area of the cathode material, which increases the effective contact area between active positive electrode materials with the electrolyte and facilitates the diffusion of lithium ions. Progress is still limited despite extensive studies. A substantial increase in capacity would make it possible to overcome obstacles to the energy density provided by a LIB. Focus on the development of next-generation electrode materials with 3D mesoporous hierarchical nanoarchitectures that offer electron channels to balance the intra- and inter-conductivity, withstand polarization during the charge–discharge process, and provide good mechanical and thermal stability is therefore needed for high-energy applications.

References

[1] https://fortunebusinessinsights.com/industry-reports/battery-energy-storage-market-100489

[2] https://iea.org/reports/energy-storage

[3] TERI 2020 *Successful Operation of Electric Bus Fleet—A Case Study of Kolkata* (New Delhi: The Energy and Resources Institute)

[4] Sheldon T L 2022 Evaluating electric vehicle policy effectiveness and equity *Ann. Rev. Resour. Econ.* **14**

[5] Abo-Khalil A G, Abdelkareem M A, Sayed E T, Maghrabie H M, Radwan A, Rezk H and Olabi A G 2022 Electric vehicle impact on energy industry, policy, technical barriers, and power systems *Int. J. Thermofluids* **13** 100134

[6] Wood D L III, Li J and Daniel C 2015 Prospects for reducing the processing cost of lithium ion batteries *J. Power Sources* **275** 234–42

[7] Winter M 2009 The solid electrolyte interphase: the most important and the least understood solid electrolyte in rechargeable Li batteries *Z. Phys. Chem.* **223** 1395–406

[8] Verma P, Maire P and Novák P 2010 A review of the features and analyses of the solid electrolyte interphase in Li-ion batteries *Electrochim. Acta* **55** 6332–41

[9] Gauthier M *et al* 2015 Electrode–electrolyte interface in Li-ion batteries: current understanding and new insights *J. Phys. Chem. Lett.* **6** 4653–72

[10] Edstroem K, Gustafsson T and Thomas J O 2004 The cathode–electrolyte interface in the Li-ion battery *Electrochim. Acta* **50** 397–403

[11] Duncan H, Abu-Lebdeh Y and Davidson I J 2010 Study of the cathode–electrolyte interface of $LiMn_{1.5}Ni_{0.5}O_4$ synthesized by a sol–gel method for Li-ion batteries *J. Electrochem. Soc.* **157** A528

[12] Liu S, Liu D, Wang S, Cai X, Qian K, Kang F and Li B 2019 Understanding the cathode electrolyte interface formation in aqueous electrolyte by scanning electrochemical microscopy *J. Mater. Chem.* A **7** 12993–96

[13] Wang Z, Lee J Z, Xin H L, Han L, Grillon N, Guy-Bouyssou D, Bouyssou E, Proust M and Meng Y S 2016 Effects of cathode electrolyte interfacial (CEI) layer on long term cycling of all-solid-state thin-film batteries *J. Power Sources* **324** 342–48

[14] Xu K 2014 Electrolytes and interphases in Li-ion batteries and beyond *Chem. Rev.* **114** 11503–618

[15] Chen X, Shen W, Vo T T, Cao Z and Kapoor A 2012 An overview of lithium-ion batteries for electric vehicles *2012 10th Int. Power and Energy Conf. (IPEC)* (Piscataway, NJ: IEEE)

[16] Whittingham M S 2004 Lithium batteries and cathode materials *Chem. Rev.* **104** 4271–302

[17] Gaur A, Sharma A L and Arya A (ed) 2021 *Energy Storage and Conversion Devices: Supercapacitors, Batteries, and Hydroelectric Cells* (Boca Raton, FL: CRC Press)

[18] Wang L, Wu Z, Zou J, Gao P, Niu X, Li H and Chen L 2019 Li-free cathode materials for high energy density lithium batteries. *Joule* **3** 2086–102

[19] Mishra A, Mehta A, Basu S, Malode S J, Shetti N P, Shukla S S, Nadagouda M N and Aminabhavi T M 2018 Electrode materials for lithium-ion batteries *Mater. Sci. Energy Technol.* **1** 182–87

[20] Nitta N, Wu F, Lee J T and Yushin G 2015 Li-ion battery materials: present and future *Mater. Today* **18** 252–64

[21] Bredow T, Heitjans P and Wilkening M 2004 Electric field gradient calculations for Li_xTiS_2 and comparison with Li 7 NMR results *Phys. Rev.* B **70** 115111

[22] Elazari R, Salitra G, Gershinsky G, Garsuch A, Panchenko A and Aurbach D 2012 Li-ion cells comprising lithiated columnar silicon film anodes, TiS_2 cathodes and fluoroethyene carbonate (FEC) as a critically important component *J. Electrochem. Soc.* **159** A1440

[23] Li L *et al* 2019 Vanadium disulfide flakes with nanolayered titanium disulfide coating as cathode materials in lithium-ion batteries *Nat. Commun.* **10** 1–10

[24] Ji Q *et al* 2017 Metallic vanadium disulfide nanosheets as a platform material for multifunctional electrode applications *Nano Lett.* **17** 4908–16

[25] Pham D T, Sambandam B, Kim S, Jo J, Kim S, Park S, Mathew V, Sun Y K, Kim K and Kim J 2018 Dandelion-shaped manganese sulfide in ether-based electrolyte for enhanced performance sodium-ion batteries *Commun. Chem.* **1** 1–14

[26] Meng Y S and Arroyo-de Dompablo, M E 2013 Recent advances in first principles computational research of cathode materials for lithium-ion batteries *Acc. Chem. Res.* **46** 1171–80

[27] Levi E and Aurbach D 2014 Lattice strains in the layered Mn, Ni and Co oxides as cathode materials in Li and Na batteries *Solid State Ion.* **264** 54–68

[28] Delmas C, Fouassier C and Hagenmuller P 1980 Structural classification and properties of the layered oxides *Physica B + C* **99** 81–5

[29] Karim A, Fosse S and Persson K A 2013 Surface structure and equilibrium particle shape of the $LiMn_2O_4$ spinel from first-principles calculations *Phys. Rev.* B **87** 075322

[30] Arroyo-de Dompablo M E and Amador U 2006 A computational investigation on the electrochemical properties of spinel-like $LiCoAsO_4$ as positive electrode for lithium-ion batteries *Solid State Sci.* **8** 916–21

[31] Tang Y, Zhang Y, Li W, Ma B and Chen X 2015 Rational material design for ultrafast rechargeable lithium-ion batteries *Chem. Soc. Rev.* **44** 5926–40

[32] Novikova S A and Yaroslavtsev A B 2017 Cathode materials based on olivine lithium iron phosphates for lithium-ion batteries *Rev. Adv. Mater. Sci.* **49** 129–39

[33] Goodenough J B 2007 Cathode materials: a personal perspective *J. Power Sources* **174** 996–1000

[34] Yuan L-X, Wang Z-H, Zhang W-X, Hu X-L, Chen J-T, Huang Y-H and Goodenough J B 2011 Development and challenges of $LiFePO_4$ cathode material for lithium-ion batteries *Energy Environ. Sci.* **4** 269–84

[35] Yang J *et al* 2016 Tuning structural stability and lithium-storage properties by d-orbital hybridization substitution in full tetrahedron Li_2FeSiO_4 nanocrystal *Nano Energy* **20** 117–25

[36] Gummow R J and He Y 2014 Recent progress in the development of Li_2MnSiO_4 cathode materials *J. Power Sources* **253** 315–31

[37] Cheng Q, He W, Zhang X, Li M and Wang L 2017 Modification of Li_2MnSiO_4 cathode materials for lithium-ion batteries: a review *J. Mater. Chem.* A **5** 10772–97

[38] Zhang T, Li D, Tao Z and Chen J 2013 Understanding electrode materials of rechargeable lithium batteries via DFT calculations *Prog. Nat. Sci.: Mater. Int.* **23** 256–72

[39] Meng Y S and Arroyo-de Dompablo M E 2009 First principles computational materials design for energy storage materials in lithium ion batteries *Energy Environ. Sci.* **2** 589–609

[40] Van der Ven A and Ceder G 2000 Lithium diffusion in layered Li_xCoO_2 *Electrochem. Solid-State Lett.* **3** 301

[41] Padhi A K, Nanjundaswamy K S and Goodenough J B 1997 Phospho-olivines as positive-electrode materials for rechargeable lithium batteries *J. Electrochem. Soc.* **144** 1188

[42] Goodenough J B, Mizushima K and Takeda T 1980 Solid-solution oxides for storage-battery electrodes *Jpn. J. Appl. Phys.* **19** 305

[43] Goodenough J B and Kim Y 2010 Challenges for rechargeable Li batteries *Chem. Mater.* **22** 587–603

[44] Manthiram A, Nazri G A and Pistoia G 2004 *Lithium Batteries: Science and Technology* (Cham: Springer) p 9

[45] Andriyevsky B, Doll K and Jacob T 2014 Electronic and transport properties of $LiCoO_2$ *Phys. Chem. Chem. Phys.* **16** 23412–20

[46] Shiraki S *et al* 2014 Fabrication of all-solid-state battery using epitaxial $LiCoO_2$ thin films *J. Power Sources* **267** 881–87

[47] Huang R, Hitosugi T, Fisher C A, Ikuhara Y H, Moriwake H, Oki H and Ikuhara Y 2012 Phase transitions in $LiCoO_2$ thin films prepared by pulsed laser deposition *Mater. Chem. Phys.* **133** 1101–7

[48] Prachařová J, Přidal J, Bludska J, Jakubec I, Vorlíček V, Malkova Z, Makris T D, Giorgi R and Jastrabık L 2002 $LiCoO_2$ thin-film cathodes grown by RF sputtering *J. Power Sources* **108** 204–12

[49] Kim W S 2004 Characteristics of $LiCoO_2$ thin film cathodes according to the annealing ambient for the post-annealing process *J. Power Sources* **134** 103–9

[50] Choi W G and Yoon S G 2004 Structural and electrical properties of $LiCoO_2$ thin-film cathodes deposited on planar and trench structures by liquid-delivery metalorganic chemical vapour deposition *J. Power Sources* **125** 236–41

[51] Balakrishnan T, Sankarasubramanian N, Kavitha A and Kathalingam A 2019 Studies on structural, optical, electrical and morphological properties of $LiCoO_2$ thin films prepared by sol–gel method *Mater. Res. Innov.* **23** 216–21

[52] Gao S, Wei W, Ma M, Qi J, Yang J, Chu S, Zhang J and Guo L 2015 Sol–gel synthesis and electrochemical properties of *c*-axis oriented $LiCoO_2$ for lithium-ion batteries *RSC Adv.* **5** 51483–88

[53] Chang S K, Kweon H J, Kim B K, Jung D Y and Kwon Y U 2002 Syntheses of $LiCoO_2$ for cathode materials of secondary batteries from reflux reactions at 130–200 C *J. Power Sources* **104** 125–31

[54] Burukhin A, Brylev O, Hany P and Churagulov B R 2002 Hydrothermal synthesis of $LiCoO_2$ for lithium rechargeable batteries *Solid State Ion.* **151** 259–63

[55] Qian X, Cheng X, Wang Z, Huang X, Guo R, Mao D, Chang C and Song W 2009 The preparation of $LiCoO_2$ nanoplates via a hydrothermal process and the investigation of their electrochemical behavior at high rates *Nanotechnology* **20** 115608

[56] Ghosh P, Mahanty S and Basu R N 2008 Effect of silver addition on the properties of combustion synthesized nanocrystalline $LiCoO_2$ *Mater. Chem. Phys.* **110** 406–10

[57] Hu G R, Liao G, Peng Z D, Xiao J, Zhang X L and Yu X Y 2004 Structure and electrochemical properties of $LiCoO_2$ synthesized by microwave heating *J. Cent. South Univer. Technol.* **11** 261–64

[58] Kim K H and Kim K B 2008 Ultrasound assisted synthesis of nano-sized lithium cobalt oxide *Ultrason. Sonochem.* **15** 1019–25

[59] Fu Z W, Ma J and Qin Q Z 2005 Nanostructured $LiCoO_2$ and $LiMn_2O_4$ fibers fabricated by a high frequency electrospinning *Solid State Ion.* **176** 1635–40

[60] Wu Q, Li W, Cheng Y and Jiang Z 2005 Homogenous $LiCoO_2$ nanoparticles prepared using surfactant P123 as template and its application to manufacturing ultra-thin-film electrode *Mater. Chem. Phys.* **91** 463–67

[61] Scott I D, Jung Y S, Cavanagh A S, Yan Y, Dillon A C, George S M and Lee S H 2011 Ultrathin coatings on nano-$LiCoO_2$ for Li-ion vehicular applications *Nano Lett.* **11** 414–18

[62] Myung S T, Kumagai N, Komaba S and Chung H T 2000 Preparation and electrochemical characterization of $LiCoO_2$ by the emulsion drying method *J. Appl. Electrochem.* **30** 1081–85

[63] Daniel C, Mohanty D, Li J and Wood D L 2014 Cathode materials review *AIP Conf. Proc.* 1597 (New York: American Institute of Physics) 26–43

[64] Rozier P and Tarascon J M 2015 Li-rich layered oxide cathodes for next-generation Li-ion batteries: chances and challenges *J. Electrochem. Soc.* **162** A2490

[65] Yano A, Shikano M, Ueda A, Sakaebe H and Ogumi Z 2016 $LiCoO_2$ degradation behavior in the high-voltage phase transition region and improved reversibility with surface coating *J. Electrochem. Soc.* **164** A6116

[66] Vetter J, Novák P, Wagner M R, Veit C, Möller K C, Besenhard J O, Winter M, Wohlfahrt-Mehrens M, Vogler C and Hammouche A 2005 Ageing mechanisms in lithium-ion batteries *J. Power Sources* **147** 269–81

[67] Sun L, Zhang Z, Hu X, Tian H, Zhang Y and Yang X 2019 Realization of Ti doping by electrostatic assembly to improve the stability of $LiCoO_2$ cycled to 4.5 V *J. Electrochem. Soc.* **166** A1793

[68] Tang W, Liu L L, Tian S, Li L, Yue Y B, Wu Y P, Guan S Y and Zhu K 2010 Nano-$LiCoO_2$ as cathode material of large capacity and high rate capability for aqueous rechargeable lithium batteries *Electrochem. Commun.* **12** 1524–26

[69] Azib T, Le Cras F and Porthault H 2015 Direct fabrication of $LiCoO_2$ thin-films in water–ethanol solutions by electrochemical–hydrothermal method *Electrochim. Acta* **160** 145–51

[70] Duffiet M, Blangero M, Cabelguen P E, Delmas C and Carlier D 2018 Influence of the initial Li/Co ratio in $LiCoO_2$ on the high-voltage phase-transitions mechanisms *J. Phys. Chem. Lett.* **9** 5334–38

[71] Jiang Y, Qin C, Yan P and Sui M 2019 Origins of capacity and voltage fading of $LiCoO_2$ upon high voltage cycling *J. Mater. Chem.* A **7** 20824–31

[72] Liang R, Yonezawa S, Kim J H and Inoue T 2018 Low-temperature synthesis of $LiCoO_2$ with eutectic of lithium precursors via the solid-state reaction method *J. Asian Ceram. Soc.* **6** 332–41

[73] Reddy M J K, Ryu S H and Shanmugharaj A M 2016 Synthesis of nanostructured lithium cobalt oxide using cherry blossom leaf templates and its electrochemical performances *Electrochim. Acta* **189** 237–44

[74] Manthiram A 2011 Materials challenges and opportunities of lithium ion batteries *J. Phys. Chem. Lett.* **2** 176–84

[75] Mukhopadhyay A and Sheldon B W 2014 Deformation and stress in electrode materials for Li-ion batteries *Prog. Mater Sci.* **63** 58–116

[76] Chen Z and Dahn J R 2004 Methods to obtain excellent capacity retention in $LiCoO_2$ cycled to 4.5 V *Electrochim. Acta* **49** 1079–90

[77] Gu R, Ma Z, Cheng T, Lyu Y, Nie A and Guo B 2018 Improved electrochemical performances of $LiCoO_2$ at elevated voltage and temperature with an *in situ* formed spinel coating layer *ACS Appl. Mater. Interfaces* **10** 31271–79

[78] Kim S, Choi S, Lee K, Yang G J, Lee S S and Kim Y 2017 Self-assembly of core–shell structures driven by low doping limit of Ti in $LiCoO_2$: first-principles thermodynamic and experimental investigation *Phys. Chem. Chem. Phys.* **19** 4104–13

[79] Jin Y, Xu S, Li Z, Xu K, Ding W, Song J, Wang H and Zhao J 2018 The influence of Ni doping on the structure and electrochemical properties of $LiCoO_2$ materials *J. Electrochem. Soc.* **165** A2267

[80] Luo D, Li G, Yu C, Yang L, Zheng J, Guan X and Li L 2012 Low-concentration donor-doped $LiCoO_2$ as a high performance cathode material for Li-ion batteries to operate between −10.4 and 45.4 °C *J. Mater. Chem.* **22** 22233–41

[81] Kim J, Kang H, Go N, Jeong S, Yim T, Jo Y N, Lee K T and Mun J 2017 Egg-shell structured $LiCoO_2$ by Cu^{2+} substitution to Li^+ sites via facile stirring in an aqueous copper (ii) nitrate solution *J. Mater. Chem.* A **5** 24892–900

[82] Er-Rami F E, Duffiet M, Hinkle S, Auvergniot J, Blangero M, Cabelguen P E, Song K, Weill F, Delmas C and Carlier D 2022 Understanding the role of Al doping of $LiCoO_2$ on the mechanisms upon cycling up to high voltages (⩾4.6 V vs Li^+/Li) *Chem. Mater.* **34** 4384–93

[83] Reddy M V *et al* 2014 Studies on bare and Mg-doped $LiCoO_2$ as a cathode material for lithium ion batteries *Electrochim. Acta* **128** 192–7

[84] Zhang J N, Li Q H, Li Q, Yu X Q and Li H 2018 Improved electrochemical performances of high voltage $LiCoO_2$ with tungsten doping *Chin. Phys.* B **27** 088202

[85] Needham S A, Wang G X, Liu H K, Drozd V A and Liu R S 2007 Synthesis and electrochemical performance of doped $LiCoO_2$ materials *J. Power Sources* **174** 828–31

[86] Yu J P, Han Z H, Hu X H, Zhan H, Zhou Y H and Liu X J 2014 The investigation of $LiCo_{1-x}Sb_xO_2$ as a promising cathode material for lithium-ion batteries *Electrochim. Acta* **121** 301–6

[87] Yoon M *et al* 2020 Unveiling nickel chemistry in stabilizing high-voltage cobalt-rich cathodes for lithium-ion batteries *Adv. Funct. Mater.* **30** 1907903

[88] Cho W, Myeong S, Kim N, Lee S, Kim Y, Kim M, Kang S J, Park N, Oh P and Cho J 2017 Critical role of cations in lithium sites on extended electrochemical reversibility of co-rich layered oxide *Adv. Mater.* **29** 1605578
[89] Liang J, Wu D, Hu M, Tian Y, Wei J and Zhou Z 2014 Could Li/Ni disorder be utilized positively? Combined experimental and computational investigation on pillar effect of Ni at Li sites on $LiCoO_2$ at high voltages *Electrochim. Acta* **146** 784–91
[90] Hirooka M, Sekiya T, Omomo Y, Yamada M, Katayama H, Okumura T, Yamada Y and Ariyoshi K 2020 Improvement of float charge durability for $LiCoO_2$ electrodes under high voltage and storage temperature by suppressing O1-phase transition *J. Power Sources* **463** 228127
[91] Manthiram A, Knight J C, Myung S T, Oh S M and Sun Y K 2016 Nickel-rich and lithium-rich layered oxide cathodes: progress and perspectives *Adv. Energy Mater.* **6** 1501010
[92] Zhang J N *et al* 2019 Trace doping of multiple elements enables stable battery cycling of $LiCoO_2$ at 4.6 V. *Nature Energy* **4** 594–603
[93] Shen B, Xu X, Liu W, Qin M and Wang W 2021 Realizing an excellent cycle and rate performance of $LiCoO_2$ at 4.55 V by Li-ion ic conductor surface modification *ACS Appl. Energy Mater.* **4** 12917–26
[94] Ellis B L, Lee K T and Nazar L F 2010 Positive electrode materials for Li-ion and Li-batteries *Chem. Mater.* **22** 691–714
[95] Kalyani P and Kalaiselvi N 2005 Various aspects of $LiNiO_2$ chemistry: a review *Sci. Technol. Adv. Mater.* **6** 689
[96] Nakamura K, Ohno H, Okamura K, Michihiro Y, Nakabayashi I and Kanashiro T 2000 On the diffusion of Li^+ defects in $LiCoO_2$ and $LiNiO_2$ *Solid State Ion.* **135** 143–47
[97] Kong F, Liang C, Longo R C, Zheng Y and Cho K 2018 Atomic-scale understanding of non-stoichiometry effects on the electrochemical performance of Ni-rich cathode materials *J. Power Sources* **378** 750–58
[98] Rao M M, Liebenow C, Jayalakshmi M, Wulff H, Guth U and Scholz F 2001 High-temperature combustion synthesis and electrochemical characterization of $LiNiO_2$, $LiCoO_2$ and $LiMn_2O_4$ for lithium-ion secondary batteries *J. Solid State Electrochem.* **5** 348–54
[99] Song M Y, Kim H U and Park H R 2014 Electrochemical properties of $LiNiO_2$ cathode after TiO_2 or ZnO addition *Ceram. Int.* **40** 4219–24
[100] Kim B H, Kim J H, Kwon I H and Song M Y 2007 Electrochemical properties of $LiNiO_2$ cathode material synthesized by the emulsion method *Ceram. Int.* **33** 837–41
[101] Sasaki T, Nonaka T, Oka H, Okuda C, Itou Y, Kondo Y, Takeuchi Y, Ukyo Y, Tatsumi K and Muto S 2009 Capacity-fading mechanisms of $LiNiO_2$-based lithium-ion batteries: I. Analysis by electrochemical and spectroscopic examination *J. Electrochem. Soc.* **156** A289
[102] Fu L J, Liu H, Li C, Wu Y P, Rahm E, Holze R and Wu H Q 2005 Electrode materials for lithium secondary batteries prepared by sol–gel methods *Prog. Mater Sci.* **50** 881–928
[103] Wu Y P, Rahm E and Holze R 2002 Effects of heteroatoms on electrochemical performance of electrode materials for lithium ion batteries *Electrochim. Acta* **47** 3491–507
[104] Kim J and Amine K 2002 A comparative study on the substitution of divalent, trivalent and tetravalent metal ions in $LiNi_{1-x}M_xO_2$ (M= Cu^{2+}, Al^{3+} and Ti^{4+}) *J. Power Sources* **104** 33–9
[105] Lee K K, Yoon W S, Kim K B, Lee K Y and Hong S T 2001 Characterization of $LiNi_{0.85}Co_{0.10}M_{0.05}O_2$ (M = Al, Fe) as a cathode material for lithium secondary batteries *J. Power Sources* **97** 308–12

[106] Jo M, Noh M, Oh P, Kim Y and Cho J 2014 A new high power $LiNi_{0.81}Co_{0.1}Al_{0.09}O_2$ cathode material for lithium-ion batteries *Adv. Energy Mater.* **4** 1301583

[107] Myung S T, Maglia F, Park K J, Yoon C S, Lamp P, Kim S J and Sun Y K 2017 Nickel-rich layered cathode materials for automotive lithium-ion batteries: achievements and perspectives *ACS Energy Lett.* **2** 196–223

[108] Wu Y (ed) 2015 *Lithium-ion Batteries: Fundamentals and Applications* vol 4 (Boca Raton, FL: CRC Press)

[109] Zhang Z and Zhang S S 2015 *Rechargeable Batteries: Materials, Technologies and New Trends* (Cham: Springer): International Publishing) 1–124

[110] Jin E M, Jin B, Jeon Y S, Park K H and Gu H B 2009 Electrochemical properties of $LiMnO_2$ for lithium polymer battery *J. Power Sources* **189** 620–23

[111] Radin M D, Hy S, Sina M, Fang C, Liu H, Vinckeviciute J, Zhang M, Whittingham M S, Meng Y S and Van der Ven A 2017 Narrowing the gap between theoretical and practical capacities in Li-ion layered oxide cathode materials *Adv. Energy Mater.* **7** 1602888

[112] Cho J, Kim T J and Park B 2002 The effect of a metal-oxide coating on the cycling behavior at 55 C in orthorhombic $LiMnO_2$ cathode materials *J. Electrochem. Soc.* **149** A288

[113] Xiao X, Wang L, Wang D, He X, Peng Q and Li Y 2009 Hydrothermal synthesis of orthorhombic $LiMnO_2$ nano-particles and $LiMnO_2$ nanorods and comparison of their electrochemical performances *Nano Res.* **2** 923–30

[114] Tong W, Chu Q, Meng Y, Wang X, Yang B, Gao J, Zhao X and Liu X 2018 Synthesis of mesoporous orthorhombic $LiMnO_2$ cathode materials via a one-step flux method for high performance lithium-ion batteries *Mater. Res. Express* **5** 065511

[115] Ji H, Yang G, Miao X and Hong A 2010 Efficient microwave hydrothermal synthesis of nanocrystalline orthorhombic $LiMnO_2$ cathodes for lithium batteries *Electrochim. Acta* **55** 3392–97

[116] Xu H, Sun J and Gao L 2013 Hydrothermal synthesis of $LiMnO_2$ microcubes for lithium ion battery application *Ionics* **19** 63–9

[117] Shen C, Xu H, Liu L, Hu H, Aguey-Zinsou K F and Wang L 2021 A one-step dynamic hydrothermal method for the synthesis of orthorhombic $LiMnO_2$/CNTs nanocomposites networks for Li-ion batteries *J. Alloys Compd.* **859** 157834

[118] Pang W K, Lee J Y, Wei Y S and Wu S H 2013 Preparation and characterization of Cr-doped $LiMnO_2$ cathode materials by Pechini's method for lithium ion batteries *Mater. Chem. Phys.* **139** 241–46

[119] Chen S, Cao F, Liu F, Xiang Q, Feng X, Liu L and Qiu G 2014 Facile hydrothermal synthesis and electrochemical properties of orthorhombic $LiMnO_2$ cathode materials for rechargeable lithium batteries *RSC Adv.* **4** 13693–703

[120] Ji H, Miao X, Wang L, Qian B and Yang G 2014 Effects of microwave-hydrothermal conditions on the purity and electrochemical performance of orthorhombic $LiMnO_2$ *ACS Sustain. Chem. Eng.* **2** 359–66

[121] Zhao H, Liu S, Liu X, Tan M, Wang Z, Cai Y and Komarneni S 2016 Orthorhombic $LiMnO_2$ nanorods as cathode materials for lithium-ion batteries: synthesis and electrochemical properties *Ceram. Int.* **42** 9319–22

[122] Hoang K and Johannes M 2016 Defect physics and chemistry in layered mixed transition metal oxide cathode materials:(Ni, Co, Mn) vs (Ni, Co, Al) *Chem. Mater.* **28** 1325–34

[123] Wang J H, Wang Y, Guo Y Z, Liu C W and Dan L L 2021 Electrochemical characterization of $AlPO_4$ coated $LiNi_{1/3}Co_{1/3}Mn_{1/3}O_2$ cathode materials for high temperature lithium battery application *Rare Met.* **40** 78–83

[124] Satyanarayana M, James J and Varadaraju U V 2017 Electrochemical performance of $LiNi_{0.4}Co_{0.2}Mn_{0.4}O_2$ prepared by different molten salt flux: $LiNO_3$–LiCl and $LiNO_3$–KNO_3 *Appl. Surf. Sci.* **418** 72–8

[125] Zhao X, Liu B, Yang J, Hou J, Wang Y and Zhu Y 2020 Synthesizing $LiNi_{0.5}Co_{0.2}Mn_{0.3}O_2$ with microsized peanut-like structure for enhanced electrochemical properties of lithium ion batteries *J. Alloys Compd.* **832** 154464

[126] Chen Y, Tang S, Deng S, Lei T, Li Y, Li W, Cao G, Zhu J and Zhang J 2019. Chemical coupling constructs amorphous silica modified $LiNi_{0.6}Co_{0.2}Mn_{0.2}O_2$ cathode materials and its electrochemical performances *J. Power Sources* **431** 8–16

[127] Li W, Li Y, Yang L, Chen Y, Guo J, Zhu J and Cao G L 2020 Enhancing high-voltage electrochemical performance of $LiNi_{0.7}Mn_{0.15}Co_{0.15}O_2$ cathode materials with SiO_2 coatings via electrostatic attraction forces method *Ionics* **26** 5393–403

[128] Liang J, Lu Y, Wang J, Liu X, Chen K, Ji W, Zhu Y and Wang D 2020 Well-ordered layered $LiNi_{0.8}Co_{0.1}Mn_{0.1}O_2$ submicron sphere with fast electrochemical kinetics for cathodic lithium storage. *J. Energy Chem.* **47** 188–95

[129] Ohzuku T and Makimura Y 2001 Layered lithium insertion material of $LiCo_{1/3}Ni_{1/3}Mn_{1/3}O_2$ for lithium-ion batteries *Chem. Lett.* **30** 642–43

[130] Chen C H, Wang C J and Hwang B J 2005 Electrochemical performance of layered Li $[Ni_xCo_{1-2x}Mn_x]O_2$ cathode materials synthesized by a sol–gel method *J. Power Sources* **146** 626–29

[131] Yabuuchi N and Ohzuku T 2003 Novel lithium insertion material of $LiCo_{1/3}Ni_{1/3}Mn_{1/3}O_2$ for advanced lithium-ion batteries *J. Power Sources* **119** 171–74

[132] Yabuuchi N, Makimura Y and Ohzuku T 2007 Solid-state chemistry and electrochemistry of $LiCo_{1/3}Ni_{1/3}Mn_{1/3}O_2$ for advanced lithium-ion batteries: III. Rechargeable capacity and cycleability *J. Electrochem. Soc.* **154** A314

[133] Wu S L, Zhang W, Song X, Shukla A K, Liu G, Battaglia V and Srinivasan V 2012 High rate capability of Li $(Ni_{1/3}Mn_{1/3}Co_{1/3})O_2$ electrode for Li-ion batteries *J. Electrochem. Soc.* **159** A438

[134] Ngala J K, Chernova N A, Ma M, Mamak M, Zavalij P Y and Whittingham M S 2004 The synthesis, characterization and electrochemical behavior of the layered $LiNi_{0.4}Mn_{0.4}Co_{0.2}O_2$ compound *J. Mater. Chem.* **14** 214–20

[135] Li Z, Chernova N A, Roppolo M, Upreti S, Petersburg C, Alamgir F M and Whittingham M S 2011 Comparative study of the capacity and rate capability of $LiNi_y Mn_yCo_{1-2y}O_2$ ($y = 0.5, 0.45, 0.4, 0.33$) *J. Electrochem. Soc.* **158** A516

[136] Ma M, Chernova N A, Toby B H, Zavalij P Y and Whittingham M S 2007 Structural and electrochemical behavior of $LiMn_{0.4}Ni_{0.4}Co_{0.2}O_2$ *J. Power Sources* **165** 517–34

[137] Koyama Y, Tanaka I, Adachi H, Makimura Y and Ohzuku T 2003 Crystal and electronic structures of superstructural Li_{1-x} $[Co_{1/3}Ni_{1/3}Mn_{1/3}]$ O2 ($0\leqslant x\leqslant 1$) *J. Power Sources* **119** 644–48

[138] Shim J H, Kim C Y, Cho S W, Missiul A, Kim J K, Ahn Y J and Lee S 2014 Effects of heat-treatment atmosphere on electrochemical performances of Ni-rich mixed-metal oxide $(LiNi_{0.80}Co_{0.15}Mn_{0.05}O_2)$ as a cathode material for lithium ion battery. *Electrochim. Acta* **138** 15–21

[139] Li D, Sasaki Y, Kobayakawa K and Sato Y 2006 Impact of cobalt substitution for manganese on the structural and electrochemical properties of $LiNi_{0.5}Mn_{0.5}O_2$ *Electrochim. Acta* **51** 3809–13

[140] Shu J, Ma R, Shao L, Shui M, Wu K, Lao M, Wang D, Long N and Ren Y 2014 In-situ x-ray diffraction study on the structural evolutions of $LiNi_{0.5}Co_{0.3}Mn_{0.2}O_2$ in different working potential windows *J. Power Sources* **245** 7–18

[141] Yang S, Wang X, Yang X, Bai Y, Liu Z, Shu H and Wei Q 2012 Determination of the chemical diffusion coefficient of lithium ions in spherical Li $[Ni_{0.5}Mn_{0.3}Co_{0.2}]O_2$ *Electrochim. Acta* **66** 88–93

[142] Konishi H, Yoshikawa M and Hirano T 2013 The effect of thermal stability for high-Ni-content layer-structured cathode materials, $LiNi_{0.8}Mn_{0.1-x}Co_{0.1}Mo_xO_2$ (x= 0, 0.02, 0.04) *J. Power Sources* **244** 23–8

[143] Bak S M, Hu E, Zhou Y, Yu X, Senanayake S D, Cho S J, Kim K B, Chung K Y, Yang X Q and Nam K W 2014 Structural changes and thermal stability of charged $LiNi_xMn_yCo_zO_2$ cathode materials studied by combined *in situ* time-resolved XRD and mass spectroscopy *ACS Appl. Mater. Interfaces* **6** 22594–601

[144] Liang L, Du K, Lu W, Peng Z, Cao Y and Hu G 2014 Synthesis and characterization of $LiNi_{0.6}Co_xMn_{0.4-x}O_2$ (x = 0.05, 0.1, 0.15, 0.2, 0.25 and 0.3) with high-electrochemical performance for lithium-ion batteries *Electrochim. Acta* **146** 207–17

[145] Ahn W, Lim S N, Jung K N, Yeon S H, Kim K B, Song H S and Shin K H 2014 Combustion-synthesized $LiNi_{0.6}Mn_{0.2}Co_{0.2}O_2$ as cathode material for lithium ion batteries *J. Alloys Compd.* **609** 143–49

[146] Liang L, Du K, Peng Z, Cao Y, Duan J, Jiang J and Hu G 2014 Co–precipitation synthesis of $Ni_{0.6}Co_{0.2}Mn_{0.2}(OH)_2$ precursor and characterization of $LiNi_{0.6}Co_{0.2}Mn_{0.2}O_2$ cathode material for secondary lithium batteries *Electrochim. Acta* **130** 82–9

[147] Yue P, Wang Z, Guo H, Wu F, He Z and Li X 2012 Effect of synthesis routes on the electrochemical performance of Li $[Ni_{0.6}Co_{0.2}Mn_{0.2}]O_2$ for lithium ion batteries *J Solid State Electrochem.* **16** 3849–54

[148] Emelina A L, Bykov M A, Kovba M L, Senyavin B M and Golubina E V 2011 Thermochemical properties of lithium cobaltate *Russ. J. Phys. Chem.* A **85** 357–63

[149] Zheng J, Kan W H and Manthiram A 2015 Role of Mn content on the electrochemical properties of nickel-rich layered $LiNi_{0.8-x}Co_{0.1}Mn_{0.1+x}O_2$ ($0.0 \leqslant x \leqslant 0.08$) cathodes for lithium-ion batteries *ACS Appl. Mater. Interfaces* **7** 6926–34

[150] Xiong X, Wang, Yue Z, Guo P, Wu H, F., Wang J and Li X 2013 Washing effects on electrochemical performance and storage characteristics of $LiNi_{0.8}Co_{0.1}Mn_{0.1}O_2$ as cathode material for lithium-ion batteries *J. Power Sources* **222** 318–25

[151] Lin F, Nordlund D, Li Y, Quan M K, Cheng L, Weng T C, Liu Y, Xin H L and Doeff M M 2016 Metal segregation in hierarchically structured cathode materials for high-energy lithium batteries *Nat. Energy* **1** 1–8

[152] Xia H, Luo Z and Xie J 2012 Nanostructured $LiMn_2O_4$ and their composites as high-performance cathodes for lithium-ion batteries *Prog. Nat. Sci.: Mater. Int.* **22** 572–84

[153] Anicete-Santos M, Gracia L, Beltrán A, Andrés J, Varela J A and Longo E 2008 Intercalation processes and diffusion paths of lithium ions in spinel-type structured $Li_{1+x}Ti_2O_4$: density functional theory study *Phys. Rev.* B **77** 085112

[154] Bhuvaneswari S, Varadaraju U V, Gopalan R and Prakash R 2019 Structural stability and superior electrochemical performance of Sc-doped $LiMn_2O_4$ spinel as cathode for lithium ion batteries *Electrochim. Acta* **301** 342–51

[155] Kim J S, Kim K, Cho W, Shin W H, Kanno R and Choi J W 2012 A truncated manganese spinel cathode for excellent power and lifetime in lithium-ion batteries *Nano Lett.* **12** 6358–65

[156] Chen B, Ben L, Yu H, Chen Y and Huang X 2018 Understanding surface structural stabilization of the high-temperature and high-voltage cycling performance of Al^{3+}-modified $LiMn_2O_4$ cathode material *ACS Appl. Mater. Interfaces* **10** 550–59

[157] Ling G W, Zhu X, He Y B, Song Q S, Li B, Li Y J, Yang Q H and Tang Z Y 2012 Structural and thermal stabilities of spinel $LiMn_2O_4$ materials under commercial power batteries cycling and abusive conditions *Int. J. Electrochem. Sci.* **7** 2455–67

[158] Lu D, Xu M, Zhou L, Garsuch A and Lucht B L 2013 Failure mechanism of graphite/ $LiNi_{0.5}Mn_{1.5}O_4$ cells at high voltage and elevated temperature *J. Electrochem. Soc.* **160** A3138

[159] Christiansen T L, Bøjesen E D, Søndergaard M, Birgisson S, Becker J and Iversen B B 2016 Crystal structure, microstructure and electrochemical properties of hydrothermally synthesised $LiMn_2O_4$ *CrystEngComm* **18** 1996–2004

[160] Thirunakaran R, Ravikumar R, Gopukumar S and Sivashanmugam A 2013 Electrochemical evaluation of dual-doped $LiMn_2O_4$ spinels synthesized via co-precipitation method as cathode material for lithium rechargeable batteries *J. Alloys Compd.* **556** 266–73

[161] Zhang C, Su J, Wang T, Yuan K, Chen C, Liu S, Huang T, Wu J, Lu H and Yu A 2018 Significant improvement on electrochemical performance of $LiMn_2O_4$ at elevated temperature by atomic layer deposition of TiO_2 nanocoating *ACS Sustain. Chem. Eng.* **6** 7890–901

[162] Yao J, Shen C, Zhang P, Gregory D H and Wang L 2012 Enhanced cycle ability of spinel $LiMn_2O_4$ by controlling the phase purity and structural strain *J. Phys. Chem. Solids* **73** 1390–95

[163] Wu H M, Tu J P, Yuan Y F, Li Y, Zhao X B and Cao G S 2005 Structural, morphological and electrochemical characteristics of spinel $LiMn_2O_4$ prepared by spray-drying method *Scr. Mater.* **52** 513–17

[164] Zhao X, Reddy M V, Liu H, Ramakrishna S, Rao G S and Chowdari B V R 2012 Nano $LiMn_2O_4$ with spherical morphology synthesized by a molten salt method as cathodes for lithium ion batteries *RSC Adv.* **2** 7462–69

[165] Okubo M, Mizuno Y, Yamada H, Kim J, Hosono E, Zhou H, Kudo T and Honma I 2010 Fast Li-ion insertion into nanosized $LiMn_2O_4$ without domain boundaries *ACS Nano* **4** 741–52

[166] Jiang Q, Liu D, Zhang H and Wang S 2015 Plasma-assisted sulfur doping of $LiMn_2O_4$ for high-performance lithium-ion batteries *J. Phys. Chem.* C **119** 28776–82

[167] Lee K S, Myung S T, Bang H, Amine K, Kim D W and Sun Y K 2009 Effect of protecting metal oxide (Co_3O_4) layer on electrochemical properties of spinel $Li_{1.1}Mn_{1.9}O_4$ as a cathode material for lithium battery applications *J. Power Sources* **189** 494–98

[168] He Z Q, Xiong L Z, Wu X M, Shang C H E N and Huang K L 2010 Synthesis and electrochemical properties of $LiMn_2O_4/Li_4Ti_5O_{12}$ composite *Trans. Nonferr. Metals Soc. Chin.* **20** s257–61

[169] Xiao H, Wang Y, Xie K, Cheng S and Cheng X 2018 High capacitance $LiMn_2O_4$ microspheres with different microstructures as cathode material for aqueous asymmetric supercapacitors *J. Alloys Compd.* **738** 25–31

[170] Cheng F, Wang H, Zhu Z, Wang Y, Zhang T, Tao Z and Chen J 2011 Porous $LiMn_2O_4$ nanorods with durable high-rate capability for rechargeable Li-ion batteries *Energy Environ. Sci.* **4** 3668–75

[171] Lee H W, Muralidharan P, Ruffo R, Mari C M, Cui Y and Kim D K 2010 Ultrathin spinel $LiMn_2O_4$ nanowires as high power cathode materials for Li-ion batteries *Nano Lett.* **10** 3852–56

[172] Xia H, Ragavendran K R, Xie J and Lu L 2012 Ultrafine $LiMn_2O_4$/carbon nanotube nanocomposite with excellent rate capability and cycling stability for lithium-ion batteries *J. Power Sources* **212** 28–34

[173] Wu S H and Lee P H 2017 Storage fading of a commercial 18650 cell comprised with NMC/LMO cathode and graphite anode *J. Power Sources* **349** 27–36

[174] Yao J *et al* 2022 A new lithium nickel oxide cathode material with a composite structure for high-performance Li-ion batteries https://doi.org/10.21203/rs.3.rs-1450650/v1

[175] Välikangas J, Laine P, Hietaniemi M, Hu T, Tynjälä P and Lassi U 2020 Precipitation and calcination of high-capacity $LiNiO_2$ cathode material for lithium-ion batteries *Appl. Sci.* **10** 8988

[176] Yoon C S, Choi M J, Jun D W, Zhang Q, Kaghazchi P, Kim K H and Sun Y K 2018 Cation ordering of Zr-doped $LiNiO_2$ cathode for lithium-ion batteries *Chem. Mater.* **30** 1808–14

[177] Tran T T B, Park E J, Kim H I, Lee S H, Jang H J and Son J T 2022 High rate performance of Lithium-ion batteries with Co-free $LiNiO_2$ cathode *Mater. Lett.* **316** 131810

[178] Ryu H H, Park G T, Yoon C S and Sun Y K 2019 Suppressing detrimental phase transitions via tungsten doping of $LiNiO_2$ cathode for next-generation lithium-ion batteries *J. Mater. Chem.* A **7** 18580–88

[179] Cao G, Zhu J, Li Y, Zhou Y, Jin Z, Xu B, Hai C and Zeng J 2020 Synthesis of a fine $LiNi_{0.88}Co_{0.09}Al_{0.03}O_2$ cathode material for lithium-ion batteries via a solvothermal route and its improved high-temperature cyclic performance *RSC Adv.* **10** 9917–23

[180] Loghavi M M, Eqra R and Mohammadi-Manesh H 2020 . Preparation and characteristics of graphene/Y_2O_3/$LiNi_{0.8}Co_{0.15}Al_{0.05}O_2$ composite for the cathode of lithium-ion battery *J. Electroanal. Chem.* **862** 113971

[181] Tsai Y T, Wu C Y and Duh J G 2021 Synthesis of Ni-rich NMC cathode material by redox-assisted deposition method for lithium ion batteries *Electrochim. Acta* **381** 138244

[182] Loghavi M M, Mohammadi-Manesh H and Eqra R 2019 $LiNi_{0.8}Co_{0.15}Al_{0.05}O_2$ coated by chromium oxide as a cathode material for lithium-ion batteries *J. Solid State Electrochem.* **23** 2569–78

[183] Vu N H, Dao V D, Vu H H T, Van Noi N, Tran D T, Ha M N and Pham T D 2021 Hydrothermal synthesis of Li_2MnO_3-stabilized $LiMnO_2$ as a cathode material for Li-ion battery *J. Nanomater.* **2021**

[184] Tian Y, Qiu Y, Liu Z, Wei X and Cao H 2020 $LiMnO_2$@rGO nanocomposites for high-performance lithium-ion battery cathodes *Nanotechnology* **32** 015402

[185] Coban A and Gungor H 2022 The carbon-based 3D-hierarchical cathode architecture for Li-ion batteries *Arab. J. Sci. Eng.* **47** 7147–55

[186] Nisa S S, Rahmawati M, Yudha C S, Nilasary H, Nursukatmo H, Oktaviano H S, Muzayanha S U and Purwanto A 2022 A fast approach to obtain layered transition-metal cathode material for rechargeable batteries *Batteries* **8** 4

[187] Zybert M, Ronduda H, Szczęsna A, Trzeciak T, Ostrowski A, Żero E, Wieczorek W, Rarog-Pilecka W and Marcinek M 2020 Different strategies of introduction of lithium ions into nickel manganese cobalt carbonate resulting in $LiNi_{0.6}Mn_{0.2}Co_{0.2}O_2$ (NMC622) cathode material for Li-ion batteries *Solid State Ion.* **348** 115273

[188] Yang Z, Wang Y, Chen X, Wu H and Zhang Y 2019 Mg^{2+} and Ti^{4+} co-doped spinel $LiMn_2O_4$ as lithium-ion battery cathode *ChemistrySelect* **4** 9583–89

[189] Wang H Q, Lai F Y, Li Y, Zhang X H, Huang Y G, Hu S J and Li Q Y 2015 Excellent stability of spinel $LiMn_2O_4$-based cathode materials for lithium-ion batteries *Electrochim. Acta* **177** 290–97

[190] Yue H, Huang X, Lv D and Yang Y 2009 Hydrothermal synthesis of $LiMn_2O_4$/C composite as a cathode for rechargeable lithium-ion battery with excellent rate capability *Electrochim. Acta* **54** 5363–67

[191] Li P, Luo S H, Wang J, Wang X, Tian Y, Li H, Wang Q, Zhang Y and Liu X 2021 Preparation and electrochemical properties of Al-F co-doped spinel $LiMn_2O_4$ single-crystal material for lithium-ion battery *Int. J. Energy Res.* **45** 21158–69

[192] Herle P S, Ellis B, Coombs N and Nazar L F 2004 Nano-network electronic conduction in iron and nickel olivine phosphates *Nat. Mater.* **3** 147–52

[193] Islam M S, Driscoll D J, Fisher C A and Slater P R 2005 Atomic-scale investigation of defects, dopants, and lithium transport in the $LiFePO_4$ olivine-type battery material *Chem. Mater.* **17** 5085–92

[194] Nishimura S I, Kobayashi G, Ohoyama K, Kanno R, Yashima M and Yamada A 2008 Experimental visualization of lithium diffusion in Li_xFePO_4 *Nat. Mater.* **7** 707–11

[195] Julien C, Mauger A, Vijh A and Zaghib K 2016 Lithium batteries *Lithium Batteries* (Cham: Springer)

[196] Wu J, Dathar G K P, Sun C, Theivanayagam M G, Applestone D, Dylla A G, Manthiram A, Henkelman G, Goodenough J B and Stevenson K J 2013 *In situ* Raman spectroscopy of $LiFePO_4$: size and morphology dependence during charge and self-discharge *Nanotechnology* **24** 424009

[197] Yang J and Tse J S 2011 Li ion diffusion mechanisms in $LiFePO_4$: an ab initio molecular dynamics study *J. Phys. Chem.* A **115** 13045–9

[198] Morgan D, Van der Ven A and Ceder G 2003 Li conductivity in Li_xMPO_4 (M = Mn, Fe, Co, Ni) olivine materials *Electrochem. Solid-State Lett.* **7** A30

[199] Amin R, Maier J, Balaya P, Chen D P and Lin C T 2008 Ionic and electronic transport in single crystalline $LiFePO_4$ grown by optical floating zone technique *Solid State Ion.* **179** 1683–87

[200] Chung S Y, Bloking J T and Chiang Y M 2002 Electronically conductive phospho-olivines as lithium storage electrodes *Nat. Mater.* **1** 123–28

[201] He W, Chen Q, Zhang T, Gao Y and Cao J 2015 Solvothermal synthesis of uniform $Li_3V_2(PO_4)_3$/C nanoparticles as cathode materials for lithium ion batteries *Micro Nano Lett.* **10** 67–70

[202] Zhang S M, Zhang J X, Xu S J, Yuan X J and He B C 2013 Li-ion diffusivity and electrochemical properties of $FePO_4$ nanoparticles acted directly as cathode materials in lithium ion rechargeable batteries *Electrochim. Acta* **88** 287–93

[203] Bai N, Chen H, Zhou W, Xiang K, Zhang Y, Li C and Lu H 2015 Preparation and electrochemical performance of $LiFePO_4$/C microspheres by a facile and novel co-precipitation *Electrochim. Acta* **167** 172–78

[204] Rosaiah P, Hussain O M, Zhu J and Qiu Y 2017 Spectroscopic and electrochemical properties of lithium-rich $LiFePO_4$ cathode synthesized by solid-state reaction *J. Electron. Mater.* **46** 4865–74

[205] Vediappan K, Guerfi A, Gariépy V, Demopoulos G P, Hovington P, Trottier J, Mauger A, Julien C M and Zaghib K 2014 Stirring effect in hydrothermal synthesis of nano C-$LiFePO_4$ *J. Power Sources* **266** 99–106

[206] Yu F, Lim S H, Zhen Y, An Y and Lin J 2014 Optimized electrochemical performance of three-dimensional porous $LiFePO_4$/C microspheres via microwave irradiation assisted synthesis *J. Power Sources* **271** 223–30

[207] Ziolkowska D A, Jasinski J B, Hamankiewicz B, Korona K P, Wu S H and Czerwinski A 2016 *In situ* XRD and TEM studies of sol–gel-based synthesis of $LiFePO_4$ *Cryst. Growth Des.* **16** 5006–13

[208] Ravnsbæk D B, Xiang K, Xing W, Borkiewicz O J, Wiaderek K M, Gionet P, Chapman K W, Chupas P J and Chiang Y M 2014 Extended solid solutions and coherent transformations in nanoscale olivine cathodes *Nano Lett.* **14** 1484–91

[209] Andre D, Kim S J, Lamp P, Lux S F, Maglia F, Paschos O and Stiaszny B 2015 Future generations of cathode materials: an automotive industry perspective *J. Mater. Chem.* A **3** 6709–32

[210] Delacourt C, Laffont L, Bouchet R, Wurm C, Leriche J B, Morcrette M, Tarascon J M and Masquelier C 2005 Toward understanding of electrical limitations (electronic, ionic) in $LiMPO_4$ (M = Fe, Mn) electrode materials *J. Electrochem. Soc.* **152** A913

[211] Okada S, Sawa S, Egashira M, Yamaki J I, Tabuchi M, Kageyama H, Konishi T and Yoshino A 2001 Cathode properties of phospho-olivine $LiMPO_4$ for lithium secondary batteries *J. Power Sources* **97** 430–32

[212] Wolfenstine J and Allen J 2004 $LiNiPO_4$–$LiCoPO_4$ solid solutions as cathodes *J. Power Sources* **136** 150–53

[213] Choi D, Li X, Henderson W A, Huang Q, Nune S K, Lemmon J P and Sprenkle V L 2016 $LiCoPO_4$ cathode from a $CoHPO_4 \cdot xH_2O$ nanoplate precursor for high voltage Li-ion batteries *Heliyon* **2** e00081

[214] Karthickprabhu S, Hirankumar G, Maheswaran A, Bella R D and Sanjeeviraja C 2014 Structural and electrical studies on Zn^{2+} doped $LiCoPO_4$ *J. Electrostat.* **72** 181–86

[215] Kreder K J III, Assat G and Manthiram A 2016 Aliovalent substitution of V^{3+} for Co^{2+} in $LiCoPO_4$ by a low-temperature microwave-assisted solvothermal process *Chem. Mater.* **28** 1847–53

[216] Ludwig J, Marino C, Haering D, Stinner C, Gasteiger H A and Nilges T 2017 Morphology-controlled microwave-assisted solvothermal synthesis of high-performance $LiCoPO_4$ as a high-voltage cathode material for Li-ion batteries *J. Power Sources* **342** 214–23

[217] Aravindan V, Gnanaraj J, Lee Y S and Madhavi S 2013 $LiMnPO_4$: a next generation cathode material for lithium-ion batteries *J. Mater. Chem.* A **1** 3518–39

[218] Fujimoto D, Lei Y, Huang Z H, Kang F and Kawamura J 2014 Synthesis and electrochemical performance of $LiMnPO_4$ by hydrothermal method *Int. J. Electrochem.* **2014** 768912

[219] Li J, Luo S, Ding X, Wang Q and He P 2018 Three-dimensional honeycomb-structural $LiAlO_2$-modified $LiMnPO_4$ composite with superior high rate capability as Li-ion battery cathodes *ACS Appl. Mater. Interfaces* **10** 10786–95

[220] Lu Q, Hutchings G S, Zhou Y, Xin H L, Zheng H and Jiao F 2014 Nanostructured flexible Mg-modified LiMnPO4 matrix as high-rate cathode materials for Li-ion batteries *J. Mater. Chem.* A **2** 6368–73

[221] Rajammal K, Sivakumar D, Duraisamy N, Ramesh K and Ramesh S 2016 Effect of sintering temperature on structural properties of $LiMnPO_4$ cathode materials obtained by sol–gel method *J. Sol–Gel Sci. Technol.* **80** 514–22
[222] Huang Q Y, Wu Z, Su J, Long Y F, Lv X Y and Wen Y X 2016 Synthesis and electrochemical performance of Ti–Fe co-doped $LiMnPO_4$/C as cathode material for lithium-ion batteries *Ceram. Int.* **42** 11348–54
[223] Zhu C, Wu Z, Xie J, Chen Z, Tu J, Cao G and Zhao X 2018 Solvothermal-assisted morphology evolution of nanostructured $LiMnPO_4$ as high-performance lithium-ion batteries cathode *J. Mater. Sci. Technol.* **34** 1544–49
[224] Dong J, Yu X, Sun Y, Liu L, Yang X and Huang X 2013 Triplite $LiFeSO_4F$ as cathode material for Li-ion batteries *J. Power Sources* **244** 716–20
[225] Kim M, Jung Y and Kang B 2015 High electrochemical performance of 3.9 V $LiFeSO_4F$ directly synthesized by a scalable solid-state reaction within 1 h *J. Mater. Chem.* A **3** 7583–90
[226] Tripathi R, Popov G, Sun X, Ryan D H and Nazar L F 2013 Ultra-rapid microwave synthesis of triplite $LiFeSO_4F$ *J. Mater. Chem.* A **1** 2990–94
[227] Kim M, Lee S and Kang B 2016 Fast-rate capable electrode material with higher energy density than $LiFePO_4$: 4.2V $LiVPO_4F$ synthesized by scalable single-step solid-state reaction *Adv. Sci.* **3** 1500366
[228] Ateba Mba J M, Masquelier C, Suard E and Croguennec L 2012 Synthesis and crystallographic study of homeotypic $LiVPO_4F$ and $LiVPO_4O$ *Chem. Mater.* **24** 1223–34
[229] Kim M, Lee S and Kang B 2017 High energy density polyanion electrode material: $LiVPO_4O_{1-x}F_x$ ($x \approx 0.25$) with tavorite structure *Chem. Mater.* **29** 4690–99
[230] Wen B *et al* 2016 Molybdenum substituted vanadyl phosphate ε-$VOPO_4$ with enhanced two-electron transfer reversibility and kinetics for lithium-ion batteries *Chem. Mater.* **28** 3159–70
[231] Wangoh L W *et al* 2016 Uniform second Li-ion intercalation in solid state ϵ-$LiVOPO_4$ *Appl. Phys. Lett.* **109** 053904
[232] Rousse G and Tarascon J M 2014 Sulfate-based polyanionic compounds for Li-ion batteries: synthesis, crystal chemistry, and electrochemistry aspects *Chem. Mater.* **26** 394–406
[233] Jin M, Gao R, Sun G, Li H, Xue X, Qu C, Li N, Zhang Y, Wang Z and Feng M 2021 Dual-function $LiFePO_4$ modified separator for low-overpotential and stable Li-S battery *J. Alloys Compd.* **873** 159798
[234] Meng Y F, Liang H J, Zhao C D, Li W H, Gu Z Y, Yu M X, Zhao B, Hou X K and Wu X L 2022 Concurrent recycling chemistry for cathode/anode in spent graphite/$LiFePO_4$ batteries: designing a unique cation/anion-co-workable dual-ion battery *J. Energy Chem.* **64** 166–71
[235] Shi M, Li R and Liu Y 2021 *In situ* preparation of $LiFePO_4$/C with unique copolymer carbon resource for superior performance lithium-ion batteries *J. Alloys Compd.* **854** 157162
[236] Liu X *et al* 2022 Electrochemical performance and behavior mechanism for Zn/$LiFePO_4$ battery in a slightly acidic aqueous electrolyte *ChemSusChem* **15** e202102631
[237] Wang C, Yuan X, Tan H, Jian S, Ma Z, Zhao J, Wang X, Chen D and Dong Y 2021 Three-dimensional carbon-coated $LiFePO_4$ cathode with improved Li-ion battery performance *Coatings* **11** 1137

[238] Park S, Oh J, Kim J M, Guccini V, Hwang T, Jeon Y, Salazar-Alvarez G and Piao Y 2020 Facile preparation of cellulose nanofiber derived carbon and reduced graphene oxide co-supported $LiFePO_4$ nanocomposite as enhanced cathode material for lithium-ion battery *Electrochim. Acta* **354** 136707

[239] Min H, Shu H, Zhou Y, Liang Q, Li X, Ren Q, Ma Z and Wang X 2020 $LiMnPO_4$ nanoplates with optimal crystal orientation *in situ* anchored on the expanded graphite for high-rate and long-life lithium ion batteries *Electrochim. Acta* **359** 136945

[240] Liang L, Sun X, Zhang J, Hou L, Sun J, Liu Y, Wang S and Yuan C 2019 *In situ* synthesis of hierarchical core double-shell Ti-doped $LiMnPO_4$@$NaTi_2(PO_4)_3$@C/3D graphene cathode with high-rate capability and long cycle life for lithium-ion batteries *Adv. Energy Mater.* **9** 1802847

[241] Vásquez F A and Calderón J A 2019 Vanadium doping of $LiMnPO_4$ cathode material: correlation between changes in the material lattice and the enhancement of the electrochemical performance *Electrochim. Acta* **325** 134930

[242] Luo S H, Sun Y, Bao S, Li J, Zhang J and Yi T F 2019 Synthesis of Er-doped $LiMnPO_4$/C by a sol-assisted hydrothermal process with superior rate capability *J. Electroanal. Chem.* **832** 196–203

[243] Li J, Luo S H, Ding X, Wang Q and He P 2018 NaCl-template assisted synthesis of 3D honeycomb-like $LiMnPO_4$/C with high rate and stable performance as lithium-ion battery cathodes *ACS Sustain. Chem. Eng.* **6** 16683–91

[244] Li J, Luo S H, Wang Q, Yan S, Feng J, Liu H, Ding X and He P 2018 Facile synthesis of carbon-$LiMnPO_4$ nanorods with hierarchical architecture as a cathode for high-performance Li-ion batteries *Electrochim. Acta* **289** 415–21

[245] Zhang J, Wei H, Cao Y, Peng C and Zhang B 2018 Hierarchical $LiMnPO_4 \cdot Li_3V_2(PO_4)_3$/C/rGO nanocomposites as superior-rate and long-life cathodes for lithium ion batteries *J. Alloys Compd.* **769** 332–39

[246] Nasir M H, Janjua N K and Santoki J 2020 Electrochemical performance of carbon modified $LiNiPO_4$ as Li-ion battery cathode: a combined experimental and theoretical study *J. Electrochem. Soc.* **167** 130526

[247] Li Y and Taniguchi I 2019 Facile synthesis of spherical nanostructured $LiCoPO_4$ particles and its electrochemical characterization for lithium batteries *Adv. Powder Technol.* **30** 1434–41

[248] Islam M, Akbar M, Ali G, Nam K W, Chung K Y and Jung H G 2020 A high voltage Li-ion full-cell battery with $MnCo_2O_4$/$LiCoPO_4$ electrodes *Ceram. Int.* **46** 26147–55

[249] Wang Y, Qiu J, Yu Z, Ming H, Li M, Zhang S and Yang Y 2018 AlF_3-modified $LiCoPO_4$ for an advanced cathode towards high energy lithium-ion battery *Ceram. Int.* **44** 1312–20

[250] Hou Y, Chang K, Li B, Tang H, Wang Z, Zou J, Yuan H, Lu Z and Chang Z 2018 Highly [010]-oriented self-assembled $LiCoPO_4$/C nanoflakes as high-performance cathode for lithium ion batteries *Nano Res.* **11** 2424–35

[251] Islam M S, Dominko R, Masquelier C, Sirisopanaporn C, Armstrong A R and Bruce P G 2011 Silicate cathodes for lithium batteries: alternatives to phosphates? *J. Mater. Chem.* **21** 9811–18

[252] Arroyo-de Dompablo M E, Armand M, Tarascon J M and Amador U 2006 On-demand design of polyoxianionic cathode materials based on electronegativity correlations: an exploration of the Li_2MSiO_4 system (M = Fe, Mn, Co, Ni) *Electrochem. Commun.* **8** 1292–98

[253] Yang J, Kang X, He D, Zheng A, Pan M and Mu S 2015 Graphene activated 3D-hierarchical flower-like Li_2FeSiO_4 for high-performance lithium-ion batteries *J. Mater. Chem.* A **3** 16567–73

[254] Sirisopanaporn C, Dominko R, Masquelier C, Armstrong A R, Mali G and Bruce P G 2011 Polymorphism in $Li_2(Fe,Mn)SiO_4$: a combined diffraction and NMR study *J. Mater. Chem.* **21** 17823–31

[255] Zhang P, Hu C H, Wu S Q, Zhu Z Z and Yang Y 2012 Structural properties and energetics of Li_2FeSiO_4 polymorphs and their delithiated products from first-principles *Phys. Chem. Chem. Phys.* **14** 7346–51

[256] Saracibar A, Van der Ven A and Arroyo-de Dompablo M E 2012 Crystal structure, energetics, and electrochemistry of Li_2FeSiO_4 polymorphs from first principles calculations *Chem. Mater.* **24** 495–503

[257] Sirisopanaporn C, Masquelier C, Bruce P G, Armstrong A R and Dominko R 2011 Dependence of Li_2FeSiO_4 electrochemistry on structure *JACS* **133** 1263–65

[258] Deng C, Zhang S, Zhao G S, Dong Z, Shang Y, Wu Y X and Zhao B D 2013 Synthesis and characterization of $(1-x)Li_2FeSiO_4{\cdot}xLi_3PO_4/C$ as a cathode material for lithium ion batteries *J. Electrochem. Soc.* **160** A1457

[259] Zhu H, Wu X, Zan L and Zhang Y 2014 Superior electrochemical capability of $Li_2FeSiO_4/C/G$ composite as cathode material for Li-ion batteries *Electrochim. Acta* **117** 34–40

[260] Kuganathan N and Islam M S 2009 Li_2MnSiO_4 lithium battery material: atomic-scale study of defects, lithium mobility, and trivalent dopants *Chem. Mater.* **21** 5196–202

[261] Duncan H, Kondamreddy A, Mercier P H, Le Page Y, Abu-Lebdeh Y, Couillard M, Whitfield P S and Davidson I J 2011 Novel Pn polymorph for Li_2MnSiO_4 and its electrochemical activity as a cathode material in Li-ion batteries *Chem. Mater.* **23** 5446–56

[262] Sun D, Wang H, Ding P, Zhou N, Huang X, Tan S and Tang Y 2013 In-situ synthesis of carbon coated Li_2MnSiO_4 nanoparticles with high rate performance *J. Power Sources* **242** 865–71

[263] Hwang C, Kim T, Shim J, Kwak K, Ok K M and Lee K K 2015 Fast ultrasound-assisted synthesis of Li_2MnSiO_4 nanoparticles for a lithium-ion battery *J. Power Sources* **294** 522–29

[264] Dong Y, Zhang W L, Wang C M, Shi T and Chen L 2015 Synthesis of La-doped Li_2MnSiO_4 nano-particle with high-capacity via polyol-assisted hydrothermal method *Electrochim. Acta* **166** 40–6

[265] Kempaiah D M, Rangappa D and Honma I 2012 Controlled synthesis of nanocrystalline Li_2MnSiO_4 particles for high capacity cathode application in lithium-ion batteries *Chem. Commun.* **48** 2698–700

[266] Aono S, Tsurudo T, Urita K and Moriguchi I 2013 Direct synthesis of novel homogeneous nanocomposites of Li_2MnSiO_4 and carbon as a potential Li-ion battery cathode material *Chem. Commun.* **49** 2939–41

[267] Zhang S, Lin Z, Ji L, Li Y, Xu G, Xue L, Li S, Lu Y, Toprakci O and Zhang X 2012 Cr-doped Li_2MnSiO_4/carbon composite nanofibers as high-energy cathodes for Li-ion batteries *J. Mater. Chem.* **22** 14661–66

[268] Gao K, Wang K and Li S 2019 Optimum synthesis and electrochemical properties of Li_2FeSiO_4 cathode for lithium ion battery prepared by a molten salt assisted method *Polyhedron* **166** 28–35

[269] Li J, Zhang M, Xue T, Zhang D, Yan Y and Li Z 2020 Hydrothermal-assisted synthesis of Li_2FeSiO_4/C composites as cathode materials for lithium-ion batteries *Int. J. Electrochem. Sci.* **15** 587–98

[270] Tang Y Q, Liu X, Huang X B, Ding X, Zhou S B and Chen Y D 2019 Synthesis and electrochemical properties of Li_2FeSiO_4/C/Ag composite as a cathode material for Li-ion battery *J. Cent. South Univ.* **26** 1443–48
[271] Pu X, Zhao G, Ding F, Cao S and Chen Z 2018 Facile synthesis of hierarchical porous Li_2FeSiO_4/C as highly stable cathode materials for lithium-ion batteries *J. Solid State Electrochem.* **22** 877–84
[272] Li L, Han E, Yang P, Zhu L and Liu Y 2018 Study on electrochemical performance of Mg-doped Li_2FeSiO_4 cathode material for Li-ion batteries *Ionics* **24** 1869–78
[273] Shree Kesavan K, Michael M S and Prabaharan S R S 2019 Facile electrochemical activity of monoclinic Li_2MnSiO_4 as potential cathode for Li-ion batteries *ACS Appl. Mater. Interfaces* **11** 28868–77
[274] Qiu S, Pu X, Ai X, Yang H, Chen Z and Cao Y 2018 Template synthesis of mesoporous Li_2MnSiO_4@C composite with improved lithium storage properties *Electrochim. Acta* **291** 124–31
[275] Cheng H M, Zhao S X, Wu X, Zhao J W, Wei L and Nan C W 2018 Synthesis and structural stability of Cr-doped Li_2MnSiO_4/C cathode materials by solid-state method *Appl. Surf. Sci.* **433** 1067–74
[276] Zhu H, Deng W, Chen L and Zhang S 2019 Nitrogen doped carbon layer of Li_2MnSiO_4 with enhanced electrochemical performance for lithium ion batteries *Electrochim. Acta* **295** 956–65
[277] Xie M, Luo R, Chen R, Wu F, Zhao T, Wang Q and Li L 2015 Template-assisted hydrothermal synthesis of Li_2MnSiO_4 as a cathode material for lithium ion batteries *ACS Appl. Mater. Interfaces* **7** 10779–84
[278] Thayumanasundaram S, Rangasamy V S, Seo J W and Locquet J P 2019 Novel strategies to improve the structural and electrochemical stability of Li_2CoSiO_4 during cycling *Solid State Ion.* **337** 161–69
[279] Le T S, Hoa T H and Truong D Q 2019 Direct synthesis of homogeneous Li_2CoSiO_4/C for enhanced ionic transport properties in Li-ion battery *J. Electroanal. Chem.* **842** 133–39
[280] Du W, Tian A, Zhang Y, Chen Z, Zhang Z, Wang D and Li J 2020 Nanostructure and its effect on electrochemical properties of polyanionic Li_2CoSiO_4 for lithium ion batteries *Nanotechnology* **31** 425602
[281] Huai L, Du W, Zhang Z, Zhang X, Zhang Z, Chen Z, Wu J, Wang D and Li J 2020 The mechanism of V-modification in Li_2CoSiO_4 cathode material for Li-ion batteries: a combined first-principles and experimental study *Electrochim. Acta* **353** 136564
[282] Du H, Zhang X, Chen Z, Wu D, Zhang Z and Li J 2018 Carbon coating and Al-doping to improve the electrochemistry of Li_2CoSiO_4 polymorphs as cathode materials for lithium-ion batteries *RSC Adv.* **8** 22813–22
[283] Zhang Z, Chen Z, Zhang X, Wu D and Li J 2018 P-doping Li_2CoSiO_4/C cathode material: a joint experimental and theoretical study *Electrochim. Acta* **264** 166–72
[284] Zou Z, Zhang S and Li S 2020 A review of the preparation and performance improvement of V_6O_{13} as a cathode material for lithium-ion batteries *Mater. Technol.* **35** 300–15
[285] Wang D, Cao L, Huang J and Wu J 2012 Synthesis and electrochemical properties of LiV_3O_8 via an improved sol–gel process *Ceram. Int.* **38** 2647–52
[286] Choi H C, Kim S B and Jung Y M 2006 Application of soft x-ray absorption spectroscopy and two-dimensional correlation spectroscopy to the electrochemical reaction in the $Li_{1+x}V_3O_8$/Li cell *J. Mol. Struct.* **799** 91–5

[287] Sun J, Jiao L, Liu L, Wei X, Yang L, Liu S, Yuan H and Wang Y 2009 Synthesis and electrochemical performance of Ti^{4+} doped LiV_3O_8 *Chin. J. Chem.* **27** 863–67

[288] Feng Y, Li Y and Hou F 2009 Preparation and electrochemical properties of Cr doped LiV_3O_8 cathode for lithium ion batteries *Mater. Lett.* **63** 1338–40

[289] Zorkipli N N M, Yahaya M Z, Salleh N A, Purwanto A, Kheawhom S, Kaus N H M and Mohamad A A 2020 Synthesis and electrochemical performance of LiV_3O_8/graphene for aqueous lithium batteries *Ionics* **26** 2277–92

[290] Song H, Liu Y, Zhang C, Liu C and Cao G 2015 Mo-doped LiV_3O_8 nanorod-assembled nanosheets as a high performance cathode material for lithium ion batteries *J. Mater. Chem.* A **3** 3547–58

[291] Feng C Q, Chew S Y, Guo Z P, Wang J Z and Liu H K 2007 An investigation of polypyrrole–LiV_3O_8 composite cathode materials for lithium-ion batteries *J. Power Sources* **174** 1095–99

[292] Gao X W, Wang J Z, Chou S L and Liu H K 2012 Synthesis and electrochemical performance of LiV_3O_8/polyaniline as cathode material for the lithium battery *J. Power Sources* **220** 47–53

[293] Idris N H, Rahman M M, Wang J Z, Chen Z X and Liu H K 2011 Synthesis and electrochemical performance of LiV_3O_8/carbon nanosheet composite as cathode material for lithium-ion batteries *Compos. Sci. Technol.* **71** 343–49

[294] Li Z, Niu B, Liu J, Li J and Kang F 2018 Rechargeable aluminum-ion battery based on MoS_2 microsphere cathode *ACS Appl. Mater. Interfaces* **10** 9451–59

[295] Zhang W J and Huang K J 2017 A review of recent progress in molybdenum disulfide-based supercapacitors and batteries *Inorg. Chem. Front.* **4** 1602–20

[296] Bhandavat R, David L and Singh G 2012 Synthesis of surface-functionalized WS_2 nanosheets and performance as Li-ion battery anodes *J. Phys. Chem. Lett.* **3** 1523–30

[297] Luo B, Fang Y, Wang B, Zhou J, Song H and Zhi L 2012 Two dimensional graphene–SnS_2 hybrids with superior rate capability for lithium ion storage *Energy Environ. Sci.* **5** 5226–30

[298] Gong Y, Yang S, Liu Z, Ma L, Vajtai R and Ajayan P M 2013 Graphene-network-backboned architectures for high-performance lithium storage *Adv. Mater.* **25** 3979–84

[299] Legagneur V, An Y, Mosbah A, Portal R, La Salle A L G, Verbaere A, Guyomard D and Piffard Y 2001 $LiMBO_3$ (M= Mn, Fe, Co): synthesis, crystal structure and lithium deinsertion/insertion properties *Solid State Ion.* **139** 37–46

[300] Tao L, Rousse G, Chotard J N, Dupont L, Bruyere S, Hanžel D, Mali G, Dominko R, Levasseur S and Masquelier C 2014 Preparation, structure and electrochemistry of $LiFeBO_3$: a cathode material for Li-ion batteries *J. Mater. Chem.* A **2** 2060–70

[301] Bo S H *et al* 2012 Degradation and (de) lithiation processes in the high capacity battery material $LiFeBO_3$ *J. Mater. Chem.* **22** 8799–809

[302] Zhao L and Li R K 2013 Study on a multifunctional crystal $LiMnBO_3$ *Mater. Res. Bull.* **48** 277–80

[303] Yang S H, Xue H and Guo S P 2021 Borates as promising electrode materials for rechargeable batteries *Coordinat. Chem. Rev.* **427** 213551

[304] Küzma M, Dominko R, Meden A, Makovec D, Bele M, Jamnik J and Gaberšček M 2009 Electrochemical activity of Li_2FeTiO_4 and Li_2MnTiO_4 as potential active materials for Li-ion batteries: a comparison with Li_2NiTiO_4 *J. Power Sources* **189** 81–8

[305] Wang W, Wang H, Liu S and Huang J 2012 Synthesis of γ-LiV_2O_5 nanorods as a high-performance cathode for Li-ion battery *J. Solid State Electrochem.* **16** 2555–61

[306] Li N, Gong H X and Qian Y T 2013 Orthorhombic γ-LiV_2O_5 as cathode materials in lithium ion batteries: synthesis and property *Chin. J. Chem. Phys.* **26** 597
[307] Wang L, Deng L, Li Y, Ren X, Mi H, Sun L, Zhang P and Gao Y 2018 Nb^{5+} doped LiV_3O_8 nanorods with extraordinary rate performance and cycling stability as cathodes for lithium-ion batteries *Electrochim. Acta* **284** 366–75
[308] Vikraman D, Hussain S, Prasanna K, Karuppasamy K, Jung J and Kim H S 2019 Facile method to synthesis hybrid phase 1T@2H $MoSe_2$ nanostructures for rechargeable lithium ion batteries *J. Electroanal. Chem.* **833** 333–39
[309] Zhou P, Collins G, Hens Z, Ryan K M, Geaney H and Singh S 2020 Colloidal WSe_2 nanocrystals as anodes for lithium-ion batteries *Nanoscale* **12** 22307–16
[310] Bai J, Zhao B, Zhou J, Si J, Fang Z, Li K, Ma H, Dai J, Zhu X and Sun Y 2019 Glucose-induced synthesis of 1T-MoS_2/C hybrid for high-rate lithium-ion batteries *Small* **15** 1805420
[311] Yeh Y Y, Chiang W H and Liu W R 2019 Synthesis of few-layer WS_2 by jet cavitation as anode material for lithium ion batteries *J. Alloys Compd.* **775** 1251–58
[312] Yoon J, Kim I T, Bae J and Hur J 2019 High-performance ZnS@graphite composites prepared through scalable high-energy ball milling as novel anodes in lithium-ion batteries *J. Ind. Eng. Chem.* **76** 258–67
[313] Fang W, Zhao H, Xie Y, Fang J, Xu J and Chen Z 2015 Facile hydrothermal synthesis of VS_2/graphene nanocomposites with superior high-rate capability as lithium-ion battery cathodes *ACS Appl. Mater. Interfaces* **7** 13044–52
[314] Cao M, Gao L, Lv X and Shen Y 2017 TiO_2-B@VS_2 heterogeneous nanowire arrays as superior anodes for lithium-ion batteries *J. Power Sources* **350** 87–93
[315] Zhu L, Lu H, Xiao F, Yao T, Liu T, Li F, Wang J, Han X, Cheng Y and Wang H 2020 Flower-like Mn/Co glycerolate-derived α-MnS/Co_9S_8/carbon heterostructures for high-performance lithium-ion batteries *ACS Appl. Energy Mater.* **3** 10215–23
[316] Chen W, Li Q, Zhang H, Wu X, Wu W and Xu M 2021 Rational synthesis of fern leaf-like FeS_2@sulfur-doped carbon as an anode for superior lithium-ion batteries *Energy Fuels* **35** 12599–609
[317] Kartick B, Srivastava S K and Mahanty S 2013 TiS_2–MWCNT hybrid as high performance anode in lithium-ion battery *J. Nanopart. Res.* **15** 1–12
[318] Son Y, Nguyen D T and Lee Y 2022 Preparation and characterization of fluorine-substituted $LiFeBO_3$ with carbon coating to enhance electrochemical performance and stability as a cathode material for Li-ion batteries *J. Power Sources* **533** 231395
[319] Zhang D, Qiao J, Dong X, Xu B, Li R and Chang C 2018 Graphene enhanced $LiFeBO_3$/C composites as cathodes for Li-ion batteries *Int. J. Electrochem. Sci.* **13** 1744–53
[320] Zhang B, Ming L, Tong H, Zhang J F, Zheng J C, Wang X W, Li H and Cheng L 2018 Ni-doping to improve the performance of $LiFeBO_3$/C cathode material for lithium-ion batteries *J. Alloys Compd.* **740** 382–88
[321] Hong B, He X, Yi H and Hu C 2020 Effect of PVP coating on $LiMnBO_3$ cathodes for Li-ion batteries *Materials* **13** 5528
[322] Karthikeyan K and Lee Y S 2014 Microwave synthesis of high rate nanostructured $LiMnBO_3$ with excellent cyclic behavior for lithium ion batteries *RSC Adv.* **4** 31851–54
[323] Hung V N, Van-Duong D and Trang T T 2020 Spinel-layered Li_2MnTiO_{4+z} cathode material for Li-ion batteries prepared by a sol-gel method *Vietnam J. Chem.* **58** 488–93
[324] Ma T, Muslim A and Su Z 2015 Microwave synthesis and electrochemical properties of lithium manganese borate as cathode for lithium ion batteries *J. Power Sources* **282** 95–9

[325] Ma R, Shao L, Wu K, Lao M, Shui M, Chen C, Wang D, Long N, Ren Y and Shu J 2013 Electrochemical behaviors of hexagonal $LiMnBO_3$ as lithium storage host material for lithium-ion batteries *Ceram. Int.* **39** 9309–17

[326] Chen W, Liu J, Zhang X, Liu S and Zhong S 2018 Synthesis and electrochemical performance of carbon-coated $LiCoBO_3$ as cathode materials for lithium-ion batteries *Ionics* **24** 1571–77

[327] Tang A, Zhong Q, Xu G and Song H 2016 Spherical $LiCoBO_3$ particles prepared via a molten salt method for lithium ion batteries *RSC Adv.* **6** 84439–44

[328] Ragupathi V, Krishnaswamy S, Raman S, Panigrahi P, Lee J and Nagarajan G S 2018 Enhanced electrochemical performance of $LiCoBO_3$ cathode material for next generation lithium-ion batteries *Appl. Surf. Sci.* **449** 421–25

[329] Le Roux B, Bourbon C, Lebedev O I, Colin J F and Pralong V 2015 Synthesis and characterization of the $LiMnBO_3$–$LiCoBO_3$ solid solution and its use as a lithium-ion cathode material *Inorg. Chem.* **54** 5273–79

[330] Yang L, Luo S, Wang Y, Zhan Y, Wang Q, Zhang Y, Liu X, Mu W and Teng F 2020 Excess capacity on compound phases of Li_2FeTiO_4 composite cathode materials synthesized by hydrothermal reaction using optional titanium sources to boost battery performance *Chin. Chem. Lett.* **31** 3200–4

[331] Hou P, Yang L, Wang Y, Wang L, Li S and Luo S H 2020 Optimize hydrothermal synthesis and electrochemical performance of Li_2FeTiO_4 composite cathode materials by using orthogonal experimental design method *Ionics* **26** 1657–62

[332] Tang Y, Liu L, Zhao H, Zhang Y, Kong L B, Gao S, Li X, Wang L and Jia D 2018 Pseudocapacitive behaviors of Li_2FeTiO_4/C hybrid porous nanotubes for novel lithium-ion battery anodes with superior performances *ACS Appl. Mater. Interfaces* **10** 20225–30

[333] Yang M, Zhao X, Yao C, Kong Y, Ma L and Shen X 2016 Nanostructured cation disordered Li_2FeTiO_4/graphene composite as high capacity cathode for lithium-ion batteries *Mater. Technol.* **31** 537–43

[334] Zhang X, Yang L, Hao F, Chen H, Yang M and Fang D 2015 Lithium-excess research of cathode material Li_2MnTiO_4 for lithium-ion batteries *Nanomaterials* **5** 1985–94

Chapter 3

Fundamentals and perspectives of lithium–sulfur batteries

The lithium–sulfur battery (LiSB) has been considered to have the most potential for next-generation electrochemical energy storage systems due to its high theoretical energy density. LiSBs have the potential to accomplish around five to six times better performance than current LiBs in terms of specific capacity and energy density, and could be a milestone in the development of energy storage technologies for battery-powered vehicles. However, various issues, including limited capacity, poor cycling stability, the polysulfide shuttle effect, the development and growth of lithium dendrites, and safety concerns have mostly prevented Li–S batteries from being used in practical applications. Each LiSB component has undergone intensive development in order to address these issues and provide good cyclic performance with high specific capacity. The key requirements for each of the components of a lithium–sulfur battery are covered in detail in the current chapter. Additionally, it introduces the reader to materials that have been frequently used for various lithium–sulfur battery components.

3.1 Introduction

Secondary batteries are now a crucial part of electronic equipment and are vitally necessary for a variety of purposes, including the storage of sustainable energy. Due to their high energy density, light weight, small volume, and long service life, LiBs are currently the technology of choice. The development of non-aqueous electrolytes, unique strategies for creating new classes of electrode materials, and advancements in packing and assembly have all contributed to the rapid growth of LiB technology during the past few decades [1–3]. All these efforts have resulted in the energy densities of LiBs being about 250 mAh g^{-1}, approaching the theoretical limits of almost all classes of electrode materials. Good cyclic behavior and power and

doi:10.1088/978-0-7503-4881-2ch3

energy density up to a certain extent has also been achieved. Nevertheless, the safety concerns, ease of recycling, and optimum cost are still under serious consideration. Also, with this performance, the LiB in its current form can hardly gain energy density beyond 300 Wh kg^{-1}, and hence cannot meet the demand of emerging high-energy applications such as electric vehicles (EVs), smart grids, and renewable energy storage [4–6]. Therefore, it is imperative to create energy storage technologies with high energy densities. Additionally, they must be built from unique materials with a minimal environmental impact, high theoretical specific capacity, and energy density. Many alternatives have emerged as an effect of consideration of this sector as a priority area of research. The continuous growth has been driven with several battery systems such as metal–air (M–air) batteries (M = Li, Zn, Mg, Al, Na, etc), lithium–sulfur batteries, potassium batteries, polymer batteries, etc [7–9].

Now it is almost confirmed that Li is certainly one of the finest options for electrochemical devices because of its light weight, least reactive nature, lowest electrochemical equivalent and reduction potential, and ability to deliver a higher gravimetric energy density. A schematic comparison of the energy densities of several rechargeable battery systems is shown in figure 3.1 [10].

Similarly, sulfur is also high on the wish-list of future energy storage device materials, as it is the sixteenth most earth-abundant element, cheaper, more environmentally friendly, with high theoretical gravimetric capacity [11, 12]. Moreover, about 70 million tons of elemental sulfur are annually produced by miners as a by-product (desulfurization of coal, hydro-desulfurization in petroleum, etc) during refining processes of minerals obtained by mining through the earth [13, 14]. Thus, designing one such energy storage technology based on lithium and sulfur can considerably provide more specific capacity and energy density.

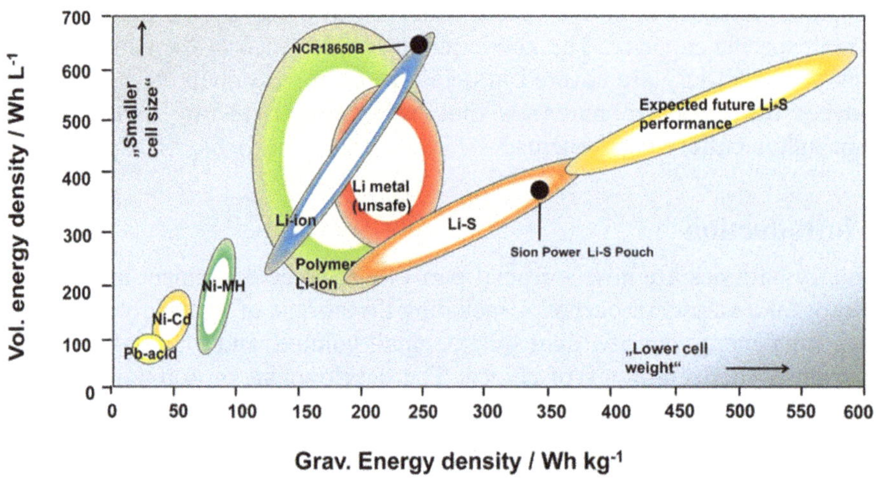

Figure 3.1. Comparison of the energy density of several electrochemical energy storage technologies. Reproduced from reference [10] John Wiley & Sons, copyright 2015 WILEY-VCH Verlag GmbH & Co. KGaA, Weinheim.

3.2 Fundamentals of lithium–sulfur batteries

Among the emerging battery technologies, lithium–sulfur batteries (LiSBs) are considered as one of the most promising choices for energy storage devices due their high theoretical specific capacity of about 1675 mAh g^{-1} at a potential of 2.2 V vs. Li/Li^+ and gravimetric energy density of about~2600 Wh kg^{-1} (or volumetric energy density ~2800 Wh L^{-1}) with sulfur as the cathode and lithium metal as the anode [11, 12, 15]. Thus, the performance of LiSBs can be about five to six times more in terms of specific capacity and energy density compared to present LiBs, which may lead a driving force to develop energy storage technologies for battery-powered vehicles.

In LiSBs, it is theoretically predicted that each sulfur (S) can accumulate two lithium ions (Li^+) and two electrons (e^-), giving rise to a large theoretical specific capacity. This can be represented as:

$$\text{Negative electrode(Li metal anode): } Li \rightarrow Li^+ + e^- \quad (3.1)$$

$$\text{Positive electrode(sulfur cathode): } S + 2e^- \rightarrow S^{2-} \quad (3.2)$$

$$\text{Overall cell reaction: } 2Li + S \rightarrow Li_2S\downarrow \quad (3.3)$$

However, during a chemical reaction elemental sulfur mostly exists as S_8, known as the crown structure or cyclooctasulfur, thus the overall redox reaction of Li/S coupling is [16, 17]:

$$S_8 + 16Li^+ + 16e^- \leftrightarrow 8Li_2S \quad (3.4)$$

However, it is well-known that the actual reaction mechanism is quite different during operation of LiSBs. It is discussed in more detail in the following section 3.2.2.

3.2.1 Cell configuration of LiSBs

A single LiSB cell comprises three main components: lithium metal foil as the anode, sulfur-hosting materials as the cathode, and an organic liquid electrolyte. A schematic of a LiSB with cell components is shown in figure 3.2.

Like LiBs, to avoid direct contact of the anode and cathode, both electrodes are immersed in a liquid electrolyte along with an ion permissible separator. In a LiSB, the energy is stored in a sulfur cathode electrode, which is further utilized during operation. According to equations (3.1) and (3.2), if oxidation of Li occurs at the anode, which is the negative electrode in LiSB, then it causes formation of Li^+ ions. If these Li^+ ions diffuse though the electrolyte towards the cathode, which is the positive electrode of LiSB, then reduction of S occurs at the cathode and hence holds the Li in the sulfur host matrix.

In a real LiSB cell, Li is used as a foil with a certain micrometer thickness, which can be used in different sizes and shapes, as Li is flexible. However, elemental sulfur is used in composites with a certain host matrix, such as carbon and its derivatives, transition metal oxides, metal chalcogenides, etc. Sulfur (both rhombic and

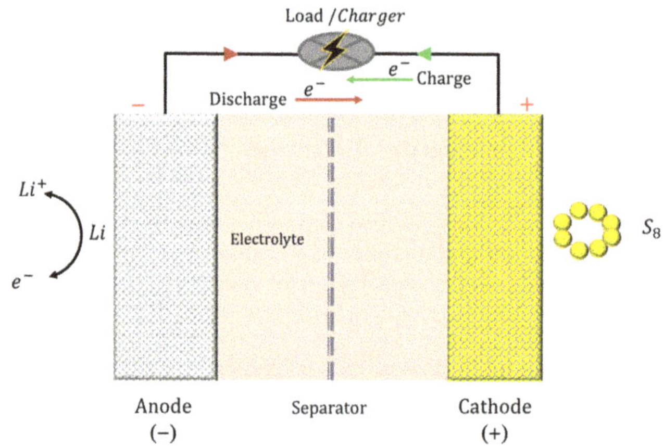

Figure 3.2. LiSB cell structure.

monoclinic) comes in the form of polyatomic molecules, among which a molecule S_8 cyclooctasulfur is most stable. At higher temperatures beyond 150 °C, the cyclo-octasulfur ring opens up due to rupturing, forming a long chain of molecules.

3.2.2 Working principle of LiSBs

In LiSB the electrodes are indirectly connected to one another via an electrolyte while being connected externally by a load. The ions move through an ion permeable separator and an ion conducting electrolyte while the electrons go through an external circuit during operation, or the charge–discharge process. By comparing the chemical potentials of the electrodes, namely the μ_A for the Li anode and the μ_C for the S cathode, the open-circuit voltage under no load conditions can be computed (figure 3.3).

The redox reaction in a cell is accomplished by μ_A and μ_C, which results in the movement of electrons and ions through the load and electrolyte, respectively. Thus, electrons go from the S cathode to the Li anode during charging, whereas they move from the Li anode to the S cathode during discharge. The essential process that takes place while a cell is operating is represented by the electrochemical reactions in equations (3.1)–(3.3).

Working mechanism

Figure 3.4 shows how a LiSB operates throughout the charge/discharge process with the passage of Li ions through the electrolyte and electrons via the external circuit.

LiSB cells' operating process often begins with prelithiation, or discharge, because the sulfur cathode accumulates in the charged state [18, 19]. Li^+ ions and electrons are produced by the oxidation at anode during discharge; the Li^+ ions then drift internally through the electrolyte, while the electrons go externally across the electrical circuit to the positive electrode. By absorbing electrons from an external circuit at the positive electrode, sulfur (S_8) is reduced, and the lithium ion then forms lithium sulfide in the electrolyte. However, in reality, the reduction of sulfur (S_8)

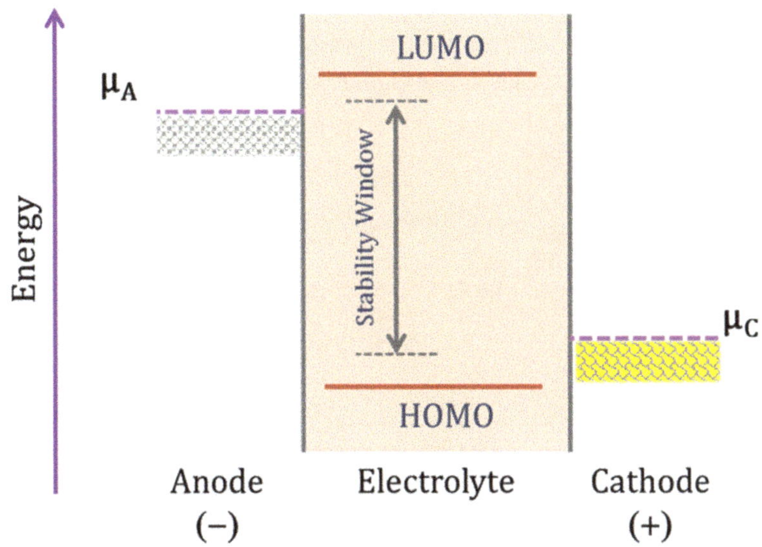

Figure 3.3. Schematic of an open-circuit energy diagram showing the redox potentials of the anode and cathode in LiSBs with respect to the HOMO-LUMO energy level of the electrolyte.

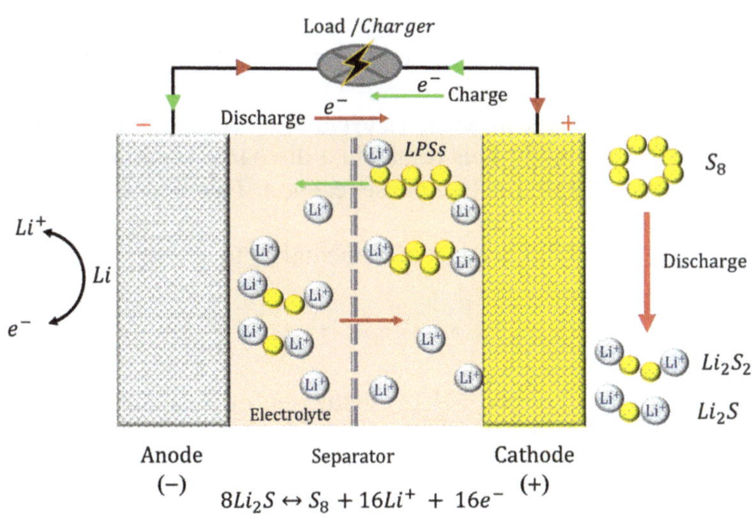

Figure 3.4. Working mechanism of LiSB during the discharge–charge process.

does not take place in a single step but rather involves the rupture of the cyclooctasulfur structure to produce high-order lithium polysulfides (LPSs), which are long homoatomic chains or homocyclic rings of varied sizes [19]. The formation of LPSs occurs with the multi-step reaction mechanism, producing intermediate polysulfides (Li_2S_n, with $1 \leqslant n \leqslant 8$) through the reduction of higher-order polysulfides to lower-order species. The reduced lower-order species may get produced simultaneously. Figure 3.5 shows a schematic representation of the overall

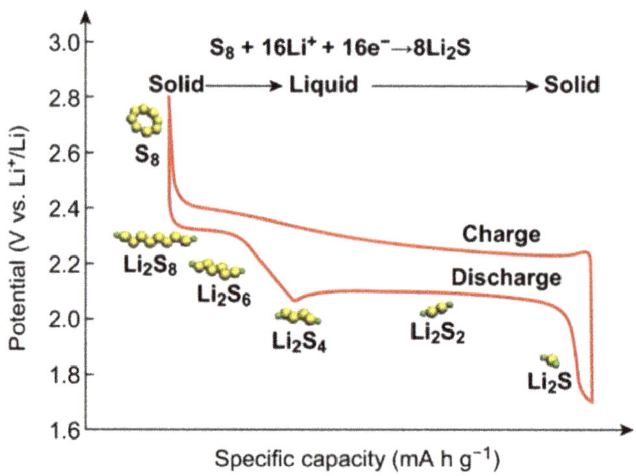

Figure 3.5. A typical charge/discharge profile for a LiSB. Reproduced from [20] John Wiley & Sons, copyright 2017 WILEY-VCH Verlag GmbH & Co. KGaA, Weinheim.

sulfur reduction, which can be systematically understood using a typical discharge and charge profile of a LiSB [20].

The two voltage plateaus in a discharge curve at 2.4 and 2.1 V are characteristic of the reduction of cyclo-sulfur to long-chain lithium polysulfides and lithium sulfide, respectively.

During an initial discharge process, the cyclo-S_8 is reduced to the long-chain high-order LPSs Li_2S_x ($4 \leqslant x \leqslant 8$). This results in a discharge plateau at high voltage of about 2.4V corresponding to the two-phase reaction. This can be represented by two steps:

Step I: At about 2.4 V: reduction of elemental S_8 to long-chain high-order LPSs (Li_2S_8)

$$S_8 + 2Li^+ + 2e^- \rightarrow Li_2S_8 \tag{3.5}$$

Step II: At 2.4 to 2.2 V (vs. Li^+/Li): reduction of long-chain high-order LPSs (Li_2S_8) to short-chain LPSs of lower order (Li_2S_6 and Li_2S_4)

$$3Li_2S_8 + 2Li^+ + 2e^- \rightarrow 4Li_2S_6 \tag{3.6}$$

$$2Li_2S_6 + 2Li^+ + 2e^- \rightarrow 3Li_2S_4 \tag{3.7}$$

With further discharge the long-chain LPSs get reduced to the low-order LPSs, Li_2S_x ($2 < x \leqslant 4$), represented by a voltage plateau at low voltage of about 2.1 V. This reduction can also be represented by two steps:

Step I: At about 2.2 V: reduction of Li_2S_4 to further low-order LPSs (Li_2S_2 or Li_2S)

$$Li_2S_4 + 2Li^+ + 2e^- \rightarrow 2Li_2S_2 \tag{3.8}$$

Step II: At 2.1 to 1.7 V (vs. Li^+/Li): reduction of Li_2S_2 to further low-order Li_2S by solid–solid reduction.

$$Li_2S_2 + 2Li^+ + 2e^- \rightarrow 2Li_2S \tag{3.9}$$

Thus, at the end of discharge, Li_2S is formed. The overall reduction process followed by the intermediate process can be represented as:

$$S_8 \rightarrow Li_2S_8 \rightarrow Li_2S_6 \rightarrow Li_2S_4 \rightarrow Li_2S_2 \rightarrow Li_2S \tag{3.10}$$

Or it can be summarized as:

$$\sum S_8 + 16\,Li^+ + 16\,e^- \rightarrow 8\,Li_2S \tag{3.11}$$

Furthermore, during a subsequent charge (de-lithiation) process, the Li_2S is converted to elemental cyclo-sulfur S_8 by oxidation. This reversible reaction is again followed by the formation of the long- and short-chain intermediate LPSs. This results in a single-charge long plateau at about 2.4 V as an overlap of two charge voltage plateaus. Typical discharge and charge curves of elemental sulfur, short-chain sulfur lithium sulfide, and Li_2S_n catholyte are shown in figure 3.6 [20].

The majority of LPSs generated during charge–discharge, nevertheless, are soluble in organic electrolytes. These dissolved LPSs can easily move between the

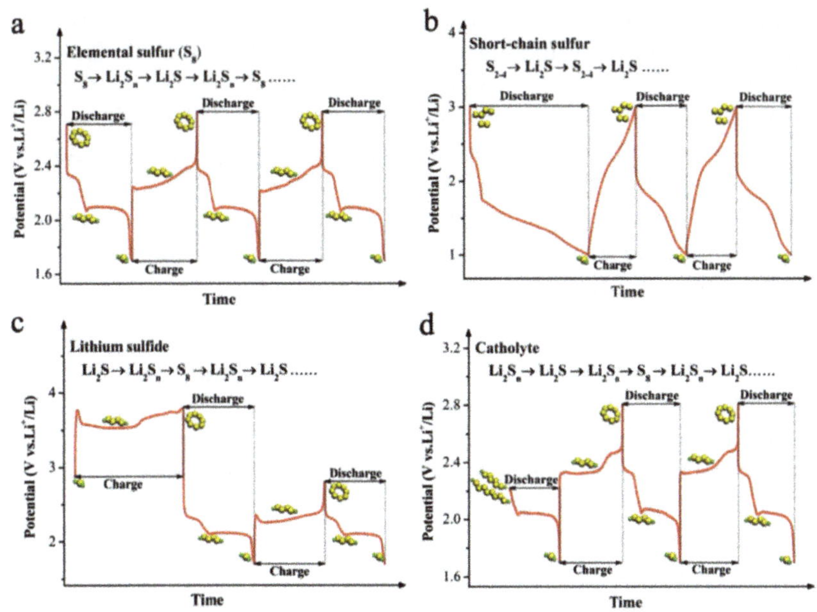

Figure 3.6. Typical discharge and charge curves of sulfur species in LiSB: (a) elemental sulfur (S_8), (b) short-chain sulfur (c) lithium sulfide, and (d) Li_2S_n catholyte. Reproduced from [20] John Wiley & Sons, copyright 2017 WILEY-VCH Verlag GmbH & Co. KGaA, Weinheim.

cathode and the anode during cell operation through the separator and electrolyte. They may react with cathode and anode materials during such a shuttling, which depletes the active material and causes capacity fading. Also, at the electrodes, a parasitic redox process arises as a result of their ongoing migration. The creation of a passivation layer by solid sulfides as a result of the interaction with lithium at the anode serves as a barrier to the transport of lithium and raises the internal cell impedance. Likewise, the reaction with the sulfur cathode causes formation of 'dead' sulfur, which results in poor stability and electrochemical performance of the LiSB cell. Such a continuous movement of LPSs between two electrodes due to the concentration gradient and the electric field while working is known as the 'shuttle effect'. This is one of the major challenges for exploration of LiSBs for high-energy applications as it causes self-discharge [21–23].

3.3 LiSB components and commonly used materials

As discussed in section 3.2, the LiSB cell has three main components: the anode, cathode, and electrolyte. The anode and cathode are supported by current collectors and are also separated internally by a separator to avoid direct contact. In addition, a suitable solvent, additives, and binders also play important roles in LiSB performance. The characteristics and key requirements of each of these components are discussed below.

The most important elements in LiSBs are lithium and sulfur. Table 3.1 displays the general properties of Li and S elements [24–28].

Cathode [29–31]

In a LiSB, elemental sulfur is taken as the cathode, i.e. a positive electrode. Sulfur is used in composite form by mixing it with a certain host matrix. To hinder the intrinsic poor electrical conductivity of sulfur, conductive carbon (carbon black) is used as an additive. For sulfur cathode electrode making, the sulfur-holding host matrix nanocomposite is coated onto an aluminum foil by making a homogenous slurry with a conducting additive and binder.

The goal of high energy density and good cyclic behavior can be achieved with a promising host matrix for sulfur cathode. The cathode host material should possess the following characteristics up to a certain moderate extent.

- High electrical conductivity;
- High thermal, structural, chemical, and electrochemical stability;
- Low cost and abundant materials;
- High surface area;
- Mesoporous network for more active electrochemical sites;
- Highly hydrophilic surface for electrolyte solvent;
- Non-corrosive;
- High chemical potential compared to HOMO of the electrolyte.

Commonly used materials for the cathode are:

- Carbon-based materials (activated carbon, carbon black, microporous carbon, mesoporous carbon) and its derivatives (CNTs, CNFs, graphene, rGO, etc);

- Conducting polymers;
- Transition metal oxides (in their binary and ternary form);
- Transition metal chalcogenides (i.e. sulfides or selenides);
- Hybrid materials in nanocomposite form.

Table 3.1. General properties of lithium and sulfur elements.

Property	Lithium	Sulfur
Element category	Metal	Non-metal
Chemical formula	Li	S_8
Atomic number	3	16
Atomic weight (amu, g mol^{-1})	6.941	32.066
Atomic radius (Å)	1.67	0.87
Van der Waals radius (Å)	1.82	1.80
Block	s-block	p-block
Group	1	16
Period	2	3
Electronic configuration	$1s^2 2s^1$ or [He] $2s^1$	$1s^2 2s^2 2p^6 3s^2 3p^4$ or [Ne] $3s^2 3p^4$
Crystal structure	Body-centered cubic	Rhombic (α-sulfur) monoclinic (β-sulfur)
Melting point (°C)	180.54	112.8 (Rhombic (α-sulfur)) 119 (monoclinic (β-sulfur))
Boiling point (°C)	1347	444.6
Density (g cm^{-3}) at 20 °C	0.534	2.07 (Rhombic (α-sulfur)) 1.96 (monoclinic (β-sulfur))
Electrical resistivity ρ (Ω m) at 20 °C	9.28×10^{-8}	1×10^{15}
Electrical conductivity σ (S m^{-1}) at 20 °C	1.1×10^{7}	1×10^{-15}
Electronegativity (Pauli scale/measurement)	0.98	2.58
Electron affinity (kJ mol^{-1})	59.6	200
First ionization potential(kJ mol^{-1})	519.9	999.3
Young's modulus	4.9 GPa	
Bulk modulus	11 GPa	7.7 GPa
Thermal expansion (μm m^{-1} K^{-1}) at 300 K	46	
Specific heat capacity (J kg^{-1} K^{-1}) at 300 K	3570	705 (Rhombic (α-sulfur))
Stable isotopes (% abundance)	^{6}Li(7.59%) and ^{7}Li (92.41%)	^{32}S (94.93%), ^{33}S(0.76%), ^{34}S (4.29%) and ^{36}S(0.02%)
Standard reduction potential E^0 (V) vs. standard hydrogen electrode (SHE) at 25 °C	$-3.040(Li^+ + e^- \rightleftharpoons Li(s))$	$-0.407(S(s) + 2e^- \rightleftharpoons S^{2-})$
Electrochemical equivalent (g A h^{-1}) (electrochemical equivalent of an element = equivalent wt/F 96 500)	0.2589	0.5981
Theoretical capacities (A h g^{-1})	3.861	1.672

Anode

This is also an important component of a LiSB for its long-term stability and high electrochemical kinetics. In principle, metallic Li foil is used as the anode material. But it suffers from formation of dendrites due to the shuttling effect of LPSs.

Electrolyte [32–36]

Electrolyte acts as the ion transport medium between the electrodes of LiSBs. Liquid organic electrolytes are most preferable to obtain better ion diffusion kinetics. The electrolyte is mostly composed of lithium salt, additives, and solvent. It should possess the following characteristics:

- High purity;
- Less viscosity;
- Low volatility;
- Effective interfacial stability;
- High thermal and chemical stability;
- High electrochemical stability;
- High ionic conductivity;
- Non-corrosive for Li metal and cathode materials;
- Lower polysulfide solubility;
- Effective interaction with cathode host matrix for active sulfur mass distribution;
- Wide range of temperature operation;
- Wide and stable potential window;
- High dielectric constant;
- Low solvated ionic radius.

Commonly used electrolytes in LiSBs are:

Solvent:

- 1,3-dioxolane (DOL);
- 1,2-dimethoxyethane (DME).

Ionic liquid (IL)-based electrolytes:

- Pyrrolidinium, piperidinium, imidazolium, and pyridinium cations and bis (trifluoromethylsulfonyl)imide (TFSI) as anions.
    - N,N-butylmethyl pyrrolidinium bis(trifluoromethylsulfonyl)imide ($Pyr_{14}TFSI$);
    - LiTFSI /$Pyr_{14}TFSI$/DOL;
    - LiTFSI in binary mixture of DOL and DME.

Separator [37–39]

It is an ion permeable porous membrane made of polyethylene or polypropylene. This is an important component of a LiSB as it prevents direct contact between the electrodes and prevents damage to the cell. It must have the following characteristics:

- High thermal stability;
- Good wettability;

- High dielectric strength and mechanical flexibility;
- Non-reactivity;
- Thin with uniform thickness;
- Chemically and electrochemically stable;
- Ability to block the shuttle of polysulfides.

Commercial materials for separators are:

- ○ Microporous polymeric membrane (polypropylene) or non-woven fabric mats or ceramic-based material;
- ○ poly(vinylidene fluoridehexafluoropropylene) (PVDF-HFP);
- ○ poly(vinylidene fluoride-cochlorotrifluoroethylene) (PVDF-CTFE);
- ○ poly(vinylidene fluoride tetrafluoroethylene) (PVDF-TFE).

Commercially available separators are:

- ○ Celgard 2500 (polypropylene based material);
- ○ Solupor 14P01A (high molecular weight polyethylene);
- ○ Gore 11367985-3 (polytetrafluoroethylene);
- ○ NKK TF40 (cellulose);
- ○ Nafion membranes.

Current collector [40–42]

Both anode and cathode material resides on a support which is a good conductor of electricity, known as a current collector. It serves as mechanical support for the Li anode and the substrate for the S host matrix and is used to collect electrons through the external circuit/load. The physical properties of an electrode like mechanical strength, porosity, film thickness and its uniformity, total mass loading, etc, vary significantly by the nature of the current collector. However, a continuous shuttling of LPSs may cause their diffusion through a flat current collector, causing poor cyclic performance.

Nowadays to avoid this, 3D porous current collectors (carbon-based or metal-based current collectors) are proposed to enhance the electrolyte/electrode interface active surface area.

Commonly used current collectors are:

- For the cathode: Al as foil, mesh, foam etched, or coated;
- For the anode: Cu as foil, mesh, foam etched, or coated.

Others explored: Ni, Ti, stainless steel foil, or mesh.

Future 3D porous current collectors: Porous carbon nano-foam and titanium carbide (MXene).

Binder [43]

To obtain proper mixing of sulfur with conducting additives, the binder plays a crucial role. The most common binders are polyvinylidene fluoride (PVDF), Poly (methyl methacrylate) (PMMA), or Polytetrafluoroethylene (PTFE). However, considering the shuttling of LPSs, polyethylene oxide (PEO) was found to be the best alternative to the conventional binders in LiSBs.

Auxiliary component: interlayers [44–48]

The intrinsic insulating nature of polysulfides limits the ion diffusion kinetics and hence causes poor overall electrochemical performance. The main cause of this is the shuttling effect due to the continuous migration of LPSs between electrodes. The separator is permeable to both ions and the LPSs, and influences the overall redox process. The practice to trap LPSs in the cathode region may overcome this issue and further results in long-term cycle stability with high sulfur utilization. This can be achieved with the help of a bifunctional interlayer between the separator and the cathode, which will work as a blocking layer for diffusion of LPSs. It is also known as a pseudo-upper current collector.

The function of the interlayer is to:

- Block LPSs from migration from the cathode to the anode;
- Stabilize the electrochemical activities within the cathode region;
- Enhance the surface conductivity of the cathode;
- Help the re-utilization of the trapped active cathode material.

The material for the functional interlayer should possess following characteristics:

- Layered 3D uniform mesoporous network;
- Ion permeability;
- Good ionic conductivity;
- Non-reactive with the electrolyte;
- Structural stability;
- Chemical stability;
- Electrochemically stable;
- Thermally stable.

The studied materials for the interlayer are:

- Microporous carbon paper;
- Mesoporous carbon paper;
- MWCNT paper;
- Interwoven CNFs;
- Porous biomaterials.

3.4 Challenges for LiSBs [21, 29–31, 49–53]

In spite of the advanced technology already in progress for the battery industry, researchers and technologists are still facing several challenges for LiSBs for their mass usage.

- The intrinsic low electrical conductivity of sulfur and low ionic conductivity of Li_2S are two of the major challenges to suppress the overall redox process in LiSBs, which needs to be overcome to improve electrochemical performance.
- Because of the higher density of sulfur (2.07 g cm^{-3}) compared to Li_2S, it suffers from structural distortion and large volume expansion during the charge–discharge process. Repeated cycling leads to pulverization of active cathode materials and leakage of liquid LPSs. It may also cause segregation

Figure 3.7. SWOT analysis of LiSBs.

of the cathode active material because of induced stress in the electrode, leading to safety issues and rapidly declined performance.

- A shuttle effect induced due to continuous dissolution of LPSs causes degradation of the electrolyte, reorganization of elemental S in the host matrix and blocking layer over the anode surface. It also induces a parasitic reaction, attributed to the prolonged charging profile. Such a continual shuttling results in rapid capacity fading, low coulombic efficiency, and poor cyclic performance.
- The deposition of low-order LPS species like Li_2S/Li_2S_2 on the surface of the anode leads to instability of the anode by surface passivation of Li. It also hinders the reduction of other LPSs, and an increase in cell impedance, causing a drop in cell voltage. The surface passivation also leads to inhomogeneous distribution of Li ions at the electrode/electrolyte interface, resulting in Li dendrite formation during repeated lithiation/de-lithiation.

It has been reported that low sulfur content and an excess amount of liquid electrolyte leads to enhanced performance. However, it is difficult for practical applications as it may cause fire, so more safety and more sealing precautions are needed. Figure 3.7 displays a SWOT (strength, weakness, opportunity, and threat) analysis of LiSB.

3.5 Efforts to overcome the challenges of LiSBs

Significant efforts have been made over each component of LiSBs to overcome these challenges and to obtain good cyclic performance with specific capacities. In this regard, the broadly identified four main areas to classify studies by several groups are:

- Optimizations of the E/S ratio for better ion diffusion kinetics and to decrease the shuttling effect;
- Electrolytes based on gel polymers and DMDS for good cyclic behavior of LiSBs;

- Rational design of porous channels in the host matrix for high sulfur loading/content;
- Host matrix with enhanced electrical conductivity for fast charge transport, better sulfur utilization, and hence good rate capabilities;
- Efficient interlayer for immobilization of LPSs.

Some studies on the fundamental understanding of processes during operation of LiSBs are also being carried out such as:

- Studies on structural and phase evolution of various poly-sulfide species and their interaction with the cathode and anode surface;
- The reaction kinetics during the growth process of Li_2S at the Li anode electrode surface as an effect of the parasitic reaction between LPSs and Li metal;
- Dynamics and interface chemistry of LPSs;
- Correlation between reaction kinetics and volume expansion of the sulfur electrode;
- Optimization of sulfur content/loading to enhance the performance without using an excess amount of electrolyte.

3.6 Opportunities for future outlook

Some of the key opportunities for enhanced cyclic and electrochemical performance of LiSBs are listed below:

- Novel design of conducting and porous host matrix with tunable pore configuration, large surface area and proportionate additive: sulfur ratio for effective sulfur loading in an active cathode are some parameters which will be a milestone in the development of LiSBs;
- Mesoporous current collectors with good structural, thermal, and chemical stability and also admirable mechanical strength and flexibility need to be developed;
- Interlayers for effective lithium-ion transfer and blocking of LPSs and strategic utilizations of sulfur need to be developed;
- Development of a polymer-based electrolyte with additives which does not allow LPSs to be dissolved in the electrolyte;
- Large specific surface area of nanostructured current collectors.

3.7 Summary

Considering their high specific capacity and energy density, lithium–sulfur batteries are seen as a sustainable candidate for advanced energy storage systems. However, the intrinsic drawback of the shuttle effect due to generated LiPSs has resulted in poor rate performance, poor specific capacity, poor cycle stabilities, and low coulombic efficiency. With the goal of obtaining long-term cycle stability with high capacity and high safety, further studies are urgently required to optimize the structure of electrode materials to improve their function in hosting sulfur, changing separators, and safeguarding Li anodes. Nanoporous carbon-based sulfur cathodes with carefully controlled and novel microstructures should be taken into consideration. Ordered and controllable structures with optimized combinations of interconnected meso- and micropores will effectively increase the sulfur loading and enhance the electrochemical properties of the cathodes. Moreover, new designs of a modified separator in Li–S batteries, optimized pore size, ion selectivity, light weight, and effective trapping of lithium polysulfides should all be taken into account. To ensure the commercial availability of Li–S batteries with long cycling stability in our daily lives, more theoretical and experimental research should be conducted.

References

[1] Masias A, Marcicki J and Paxton W A 2021 Opportunities and challenges of lithium ion batteries in automotive applications *ACS Energy Lett.* **6** 621–30

[2] Goodenough J B and Kim Y 2010 Challenges for rechargeable Li batteries *Chem. Mater.* **22** 587–603

[3] Dunn B, Kamath H and Tarascon J M 2011 Electrical energy storage for the grid: a battery of choices *Science* **334** 928–35

[4] Blomgren G E 2016 The development and future of lithium ion batteries *J. Electrochem. Soc.* **164** A5019

[5] Ma J *et al* 2021 The 2021 battery technology roadmap *J. Phys. D: Appl. Phys.* **54** 183001

[6] Verma J and Kumar D 2021 Metal-ion batteries for electric vehicles: current state of the technology, issues and future perspectives *Nanoscale Adv.* **3** 3384–94

[7] Martins L S, Guimarães L F, Junior A B B, Tenório J A S and Espinosa D C R 2021 Electric car battery: an overview on global demand, recycling and future approaches towards sustainability *J. Environ. Manage.* **295** 113091

[8] Shah R, Mittal V, Matsil E and Rosenkranz A 2021 Magnesium-ion batteries for electric vehicles: current trends and future perspectives. *Adv. Mech. Eng.* **13** 1–9

[9] Wang B, Ang E H, Yang Y, Zhang Y, Ye M, Liu Q and Li C C 2021 Post-lithium-ion battery era: recent advances in rechargeable potassium-ion batteries *Chem.–A Eur. J.* **27** 512–36

[10] Hagen M, Hanselmann D, Ahlbrecht K, Maça R, Gerber D and Tübke J 2015 Lithium–sulfur cells: the gap between the state-of-the-art and the requirements for high energy battery cells *Adv. Energy Mater.* **5** 1401986

[11] Li Y and Guo S 2021 Material design and structure optimization for rechargeable lithium-sulfur batteries *Matter* **4** 1142–88

[12] Shen X *et al* 2021 Advanced electrode materials in lithium batteries: retrospect and prospect *Energy Mater. Adv.* **2021** 1205324

[13] Chung W J *et al* 2013 The use of elemental sulfur as an alternative feedstock for polymeric materials *Nat. Chem.* **5** 518–24

[14] Etacheri V, Marom R, Elazari R, Salitra G and Aurbach D 2011 Challenges in the development of advanced Li-ion batteries: a review *Energy Environ. Sci.* **4** 3243–62

[15] Dong S, Liu H, Hu Y and Chong S 2022 Cathode materials for rechargeable lithium-sulfur batteries: current progress and future prospects *ChemElectroChem* **9** e202101564

[16] Yu S H, Huang X, Schwarz K, Huang R, Arias T A, Brock J D and Abruña H D 2018 Direct visualization of sulfur cathodes: new insights into Li–S batteries via operando X-ray based methods *Energy Environ. Sci.* **11** 202–10

[17] Song J, Gordin M L, Xu T, Chen S, Yu Z, Sohn H, Lu J, Ren Y, Duan Y and Wang D 2015 Strong lithium polysulfide chemisorption on electroactive sites of nitrogen-doped carbon composites for high-performance lithium–sulfur battery cathodes *Angew. Chem.* **127** 4399–403

[18] Lim W G, Kim S, Jo C and Lee J 2019 A comprehensive review of materials with catalytic effects in Li–S batteries: enhanced redox kinetics *Angew. Chem.* **131** 18920–31

[19] Manthiram A, Fu Y, Chung S H, Zu C and Su Y S 2014 Rechargeable lithium–sulfur batteries *Chem. Rev.* **114** 11751–87

[20] Fang R, Zhao S, Sun Z, Wang D W, Cheng H M and Li F 2017 More reliable lithium-sulfur batteries: status, solutions and prospects *Adv. Mater.* **29** 1606823

[21] Busche M R, Adelhelm P, Sommer H, Schneider H, Leitner K and Janek J 2014 Systematical electrochemical study on the parasitic shuttle-effect in lithium-sulfur-cells at different temperatures and different rates *J. Power Sources* **259** 289–99

[22] Hou L-P, Zhang X-Q, Li B-Q and Zhang Q 2021 Challenges and promises of lithium metal anode by soluble polysulfides in practical lithium–sulfur batteries *Mater. Today* **45** 62–76

[23] Mikhaylik Y V and Akridge J R 2004 Polysulfide shuttle study in the Li/S battery system *J. Electrochem. Soc.* **151** A1969–76

[24] https://periodictable.com/Properties

[25] https://chem.libretexts.org

[26] Milazzo G, Caroli S and Sharma V K 1978 *Tables of Standard Electrode Potentials* (London: Wiley)

[27] Suslick K S 1998 *Kirk-Othmer Encyclopedia of Chemical Technology* vol 26 (New York: Wiley) 517–41

[28] Lewis R A 2016 *Hawley's Condensed Chemical Dictionary.* (New York: Wiley)

[29] Peng H J and Zhang Q 2015 Designing host materials for sulfur cathodes: from physical confinement to surface chemistry *Angew. Chem. Int. Ed.* **54** 11018–20

[30] Yang Y, Zheng G and Cui Y 2013 Nanostructured sulfur cathodes *Chem. Soc. Rev.* **42** 3018–32

[31] Liu Z, Zheng X, Yuan N Y and Ding J N 2017 A novel device structure for a low-cost Li–S battery *J. Mater. Chem.* A **5** 942–6

[32] Zheng J, Gu M, Chen H, Meduri P, Engelhard M H, Zhang J G, Liu J and Xiao J 2013 Ionic liquid-enhanced solid state electrolyte interface (SEI) for lithium–sulfur batteries *J. Mater. Chem.* A **1** 8464–70

[33] Wang L and Byon H R 2013 N-Methyl-N-propylpiperidinium bis (trifluoromethanesulfonyl) imide-based organic electrolyte for high performance lithium–sulfur batteries *J. Power Sources* **236** 207–14

[34] Wang J, Chew S Y, Zhao Z W, Ashraf S, Wexler D, Chen J, Ng S H, Chou S L and Liu H K 2008 Sulfur–mesoporous carbon composites in conjunction with a novel ionic liquid electrolyte for lithium rechargeable batteries *Carbon* **46** 229–35

[35] Manan N S, Aldous L, Alias Y, Murray P, Yellowlees L J, Lagunas M C and Hardacre C 2011 Electrochemistry of sulfur and polysulfides in ionic liquids *J. Phys. Chem.* B **115** 13873–9

[36] Park J W, Ueno K, Tachikawa N, Dokko K and Watanabe M 2013 Ionic liquid electrolytes for lithium–sulfur batteries *J. Phys. Chem.* C **117** 20531–41

[37] Yuan C, Yang X, Zeng P, Mao J, Dai K, Zhang L and Sun X 2021 Recent progress of functional separators with catalytic effects for high-performance lithium-sulfur batteries *Nano Energy* **84** 105928

[38] Zhang Y, Wang Y, Luo R, Yang Y, Lu Y, Guo Y, Liu X, Cao S, Kim J K and Luo Y 2020 A 3D porous FeP/rGO modulated separator as a dual-function polysulfide barrier for high-performance lithium sulfur batteries *Nanoscale Horiz.* **5** 530–40
[39] Hao Z, Yuan L, Li Z, Liu J, Xiang J, Wu C, Zeng R and Huang Y 2016 High performance lithium-sulfur batteries with a facile and effective dual functional separator *Electrochim. Acta* **200** 197–203
[40] Kong L, Peng H J, Huang J Q and Zhang Q 2017 Review of nanostructured current collectors in lithium–sulfur batteries *Nano Res.* **10** 4027–54
[41] Zhang Y, Bakenov Z, Zhao Y, Konarov A, Wang Q and Chen P 2014 Three-dimensional carbon fiber as current collector for lithium/sulfur batteries *Ionics* **20** 803–8
[42] Chung S H and Manthiram A 2013 Lithium–sulfur batteries with superior cycle stability by employing porous current collectors *Electrochim. Acta* **107** 569–76
[43] Yuan H, Huang J Q, Peng H J, Titirici M M, Xiang R, Chen R, Liu Q and Zhang Q 2018 A review of functional binders in lithium–sulfur batteries *Adv. Energy Mater.* **8** 1802107
[44] Fan L, Li M, Li X, Xiao W, Chen Z and Lu J 2019 Interlayer material selection for lithium-sulfur batteries *Joule* **3** 361–86
[45] Chen L, Yu H, Li W, Dirican M, Liu Y and Zhang X 2020 Interlayer design based on carbon materials for lithium–sulfur batteries: a review *J. Mater. Chem.* A **8** 10709–35
[46] Su Y S and Manthiram A 2012 Lithium–sulfur batteries with a microporous carbon paper as a bifunctional interlayer *Nat. Commun.* **3** 1–6
[47] Huang J Q, Xu Z L, Abouali S, Garakani M A and Kim J K 2016 Porous graphene oxide/carbon nanotube hybrid films as interlayer for lithium-sulfur batteries *Carbon* **99** 624–32
[48] Wu K, Hu Y, Cheng Z, Pan P, Jiang L, Mao J, Ni C, Gu X and Wang Z 2019 Carbonized regenerated silk nanofiber as multifunctional interlayer for high-performance lithium-sulfur batteries *J. Membr. Sci.* **592** 117349
[49] Gupta A and Manthiram A 2022 Principles and challenges of lithium–sulfur batteries *Advances in Rechargeable Lithium–Sulfur Batteries* (Cham: Springer) 1–18
[50] Manthiram A, Fu Y and Su Y S 2013 Challenges and prospects of lithium–sulfur batteries *Acc. Chem. Res.* **46** 1125–34
[51] Yuan H, Peng H J, Huang J Q and Zhang Q 2019 Sulfur redox reactions at working interfaces in lithium–sulfur batteries: a perspective *Adv. Mater. Interfaces* **6** 1802046
[52] Li T, Bai X, Gulzar U, Bai Y J, Capiglia C, Deng W, Zhou X, Liu Z, Feng Z and Proietti Z R 2019 A comprehensive understanding of lithium–sulfur battery technology *Adv. Funct. Mater.* **29** 1901730
[53] Zhao M, Li B Q, Zhang X Q, Huang J Q and Zhang Q 2020 A perspective toward practical lithium–sulfur batteries *ACS Cent. Sci.* **6** 1095–104

Chapter 4

Host materials for sulfur cathodes in lithium–sulfur batteries

In order to address the issue of the shuttling effect initiated by the diffusion and migration of lithium polysulfides (LPSs), the sulfur cathode – one of the most important Li–S battery components has received a lot of attention. It plays a crucial role in electrochemical reactions and have significant impact on the capacity retention, coulombic efficiency and long cycling performance of LiSB. The prospective host matrix for the sulfur cathode can help to meet the objectives of high energy density and good cyclic behavior. The electrochemical performance, LPS dissolution, redox reactions at the electrolyte–electrode interface, and sulfur loading/content in the cathode material are all determined by the sulfur host matrix. Several classes of materials such as carbonaceous materials, conducting polymers, metal oxides, metal chalcogenides, MOFs, MXene, etc, have been explored as the sulfur host matrix in LiSBs. During the past decade, tremendous efforts have been devoted to the design and synthesis of electrode materials that can trap or capture polysulfide significantly. Notable achievements through a wide variety of strategies for rational design of sulfur host material classes have been obtained. Benefiting from their tunable structural parameters and hollow porous network, remarkable enhancements in the development of high-performance LiSBs have been achieved. The present chapter discusses the performance of LiSBs with different sulfur host materials, with a focus on sulfur hosts based on carbon and its many derivatives, metal oxides, and metal sulfides. The efforts made to enhance the performance of common materials from each of these classes have been widely discussed.

4.1 Introduction

Significant progress in the last decade has proved that LiSB technology can be a promising alternative to all of the existing secondary batteries because of its high

doi:10.1088/978-0-7503-4881-2ch4

theoretical specific capacity and energy density values. However, the actual electrochemical performance is still far from theoretical prediction. Continuous efforts have been made to resolve the issues which obstruct the practical electrochemical performance of LiSBs like volume expansion of sulfur, low electrical conductivity, sulfur loss during cycling, shuttling effect, dissolution of LPSs, Li dendrite formation, etc. Research on almost all of the components of LiSBs including allied ones such as current collectors, binders, additives, and solvents for electrolytes, has been done up to a certain extent. To obtain the theoretically expected values, a number of combinations and permutations of each of these have been investigated. From the research that has been done, it is clear that the cathode has attracted a lot of attention because of its important role in electrochemical processes and significant contributions to capacity, coulombic efficiency, and extended cycle life [1–3]. To date, progress has been made through a wide range of methods for the rational design of cathode materials and by tackling the main, long-standing technical issues with LiSBs. The studies on several combinations, designs, and electrochemical functionality of novel composite materials acting as sulfur hosts are explored in this chapter. Current research developments on S cathodes are outlined along with their prospects for the future.

4.2 Prerequisite of cathodes in LiSBs

The cathode is an important component of LiSBs, and its key requirements are discussed in chapter 3 section 3.3.

The mixture of elemental sulfur, a conducting additive, a host matrix, and a binder makes up the active composite of the cathode electrode. Among these, the host matrix decides the electrical conductivity, chemical thermal stability, electrochemical performance, dissolution of LPSs, redox reactions at the electrolyte–electrode interface, and the amount of sulfur loading/content in the cathode material. Several classes of materials such as carbonaceous materials, conducting polymers, metal oxides, metal chalcogenides, etc, have been explored as a sulfur host matrix in LiSBs. Among all of these, activated carbon and carbon black are well accepted by the industry as conductive additives, much like LiB, due to their easy processing, low cost, high abundance and good electrical conductivity. However, problems like the ion and electron transport kinetics, shuttling effect due to LPSs, pulverization of structure and volume, and effective sulfur content have not yet been sufficiently resolved. The following is a list of the general characteristics for the host matrix in LiSBs that arose from an invoked approach toward the cathode electrode from both an experimental and theoretical point of view [1–4].

- High electrical conductivity;
- Good structural stability;
- Mesoporous continuous network;
- Moderate density of macro- and micropores in the network;
- Physical or chemical trapping of lithium polysulfide.

4.3 Anchoring approaches toward sulfur species

The dissolution of the PS species is the most common obstacle for efficient lithiation and de-lithiation processes. It not only causes the shuttling effect due to diffusion and migration of LPSs but also volumetric pulverization of the cathode, which may result in active material loss, and limit the utilization of active material from the electrodes. The design and construction of novel sulfur host materials that can trap or capture PSs is one of the most successful approaches to obtain enhanced performance of LiSBs in terms of high reversible capacity and cycling stability. The structural design of the cathode host materials is intended in a way that they can efficiently anchor the LPSs [1, 5–7].

In broad view, the strategies to confine sulfur species are of two types:

- **Physical confinement;**
- **Chemical binding/anchoring** (through chemical bonding between sulfur species and the functional group of the host material).

The physical confinement is also known as the sulfur fixing approach for anchoring of LPSs, while chemical anchoring is the sulfur capturing approach for the trapping of LPSs [5–7]. Figure 4.1 illustrates the anchoring strategies of host materials towards sulfur species.

4.3.1 S fixing

Physical confinement means spatial confinement of sulfur species inside the structure of the host material. This can be done by physical adsorption into the host matrix or the coating of an external layer [2–5]. Thus, for designing a sulfur host structure with a porous network that can act as barrier for dissolution of LPSs, there are two categories of material structure which can succeed with this aim [8, 9].

These are:

- Porous structures;
- Layered structures.

Porous structures

Host materials with micro, meso, and macropores and large specific surface area are able to effectively accommodate elemental sulfur and confine LPSs into their porous structural configurations. Such a physical confinement is achieved by encapsulation of sulfur species within the pores. The pore wall acts as a barrier for diffusion of LPSs [8, 9]. However, the microporous network limits the sulfur loading/content in the cathode, leading to low energy density and specific capacity [10]. Various carbon-based materials such as porous carbon, carbon nanofibers, reduced graphene oxide, carbon nanotubes, porous organic polymers, etc, have been studied for physical confinement [1, 9].

Recently, 'sulfiphilic' surface chemistry is an emerging approach for physical confinement of polysulfide. It causes uniform geospatial distribution of sulfur species at the cathode surface. This can be done by activating affinity of LPSs to the host material by introducing polar functional groups or heteroatoms in it [9, 11]. In general,

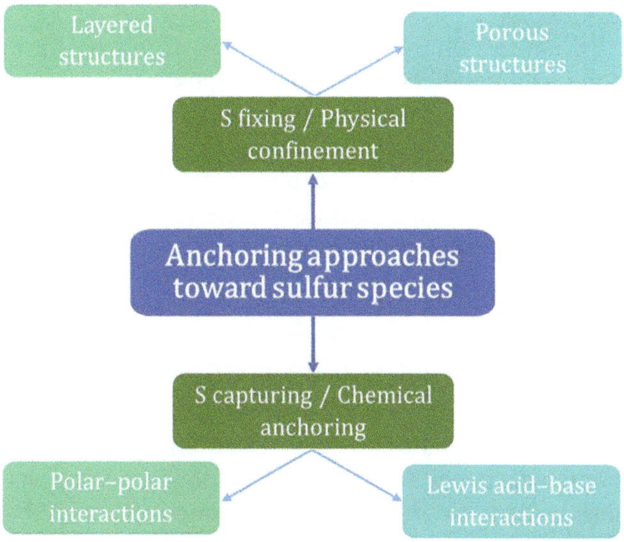

Figure 4.1. Anchoring strategies of host materials towards sulfur species.

oxygen-containing functional groups are used for the 'sulfiphilic' polar surface of the host material [12]. However, this approach causes a decrease in electrical conductivity of the cathode significantly. Thus, a host material with both sulfiphilic and conductive features can provide effective physical confinement of LPSs [13, 14].

Layered structures

A thin material layer over the sulfur electrode acts as an effective barrier for diffusion of polysulfide. This can be achieved by two approaches: either coating an electrode surface by a thin layer or by preparing a core–shell or yolk-core–shell structure. Both approaches prevent the loss of sulfur by hindering the movement of the soluble polysulfide species during the lithiation/de-lithiation processes. Two-dimensional materials like graphene are the most preferable for coating, as they offer a high specific surface area and good electrical conductivity [1, 2, 5, 13].

4.3.2 S capturing

Another approach for anchoring sulfur species is chemical adsorption or bonding between sulfur species and the functional group of the host material. This approach has been reported as the most convenient and effective, as it does not depends on the structure of the host matrix [13–17]. The strong chemical interaction of sulfur and the host material can be achieved by four ways:

- Polar–polar interactions;
- Lewis acid–base interactions;
- Redox interactions;
- Covalent fixing of sulfur species.

Polar–polar interactions
LPSs are polar in nature, thus they can interact strongly with a polar surface. Metal oxides stand as a good choice for this purpose as the metal forms a strong bond with sulfur and oxygen with lithium in LPSs. However, the low electrical conductivity of metal oxides results in low capacity and poor cyclic behavior. Conductive nonpolar carbon materials can be converted to polar by doping them with suitable metal cations. Such heteroatom-doped carbon materials are emerging as the best materials for chemical anchoring of sulfur species through polar–polar interaction [18, 19].

Lewis acid–base interactions
Most LPSs have an orbital occupied by lone electron pairs, and thus can be a Lewis base. The Lewis acids have unoccupied orbitals so they can accept the electrons from Lewis base materials. Thus, the anchoring of LPSs is possible through interaction with Lewis acids by sharing lone pair electrons. The emerging classes of materials such as metal–organic frameworks (MOFs) and MXene nanosheets as a host material for active cathodes, can significantly trap the LPSs through chemical bonding, leading to LiSBs with enhanced electrochemical performance [20].

Two other approaches for chemical adsorption have been explored, based on redox interactions and covalent fixing of sulfur species at the surface of host materials. For the first one, the redox potential window of the host must be above that of the LPSs, while in the case of the latter, only low-order short-chain polysulfides can be anchored onto the surface. For such an anchoring in the LiSB cathode, some metal oxides and conducting polymers as host materials have been investigated [21, 22].

Thus, in summary, host materials with an optimized mesoporous network and pore volume, large specific surface area, good electrical conductivity, and sufficient bonding sites can provide effective physical confinement as well as chemical adsorption of sulfur species. They will not only help to suppress the shuttling effect due to dissolution of LPSs but also improve sulfur loading/content significantly.

4.4 Cathode materials for LiSBs

Appropriate design, construction, and development of novel cathode materials possessing large specific surface area, 3D porous structures, and good electrical conductivity will be the most promising candidates for LiSBs with excellent rate performance and super high energy density. Aiming towards this, various sulfur host materials have been explored extensively. Detailed analysis of the studies will provide unprecedented opportunities to understand the key performance governing parameters and future outlook [1, 9, 14]. Different classes of materials are used as sulfur hosts in LiSBs, as shown in figure 4.2.

The following sections discus development strategies with enough electrochemical performance details of sulfur host materials, in particular carbon, metal oxide, and metal chalcogenides.

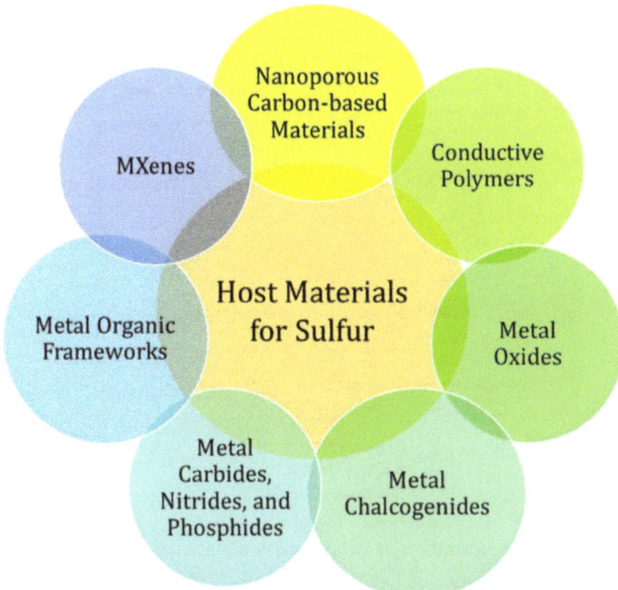

Figure 4.2. Schematic representation of the sulfur host materials in LiSBs.

4.5 Carbon-based host materials for sulfur cathodes

Since the first demonstration of Li ion based energy storage technology, carbon materials have been paid great attention as conducting agents in the electrode. Likewise, many carbon-based structures have been investigated as promising host materials for the cathode of LiSBs. Carbon materials have been considered as the most versatile class of materials because of their abundance, low cost, tunable properties, easy synthesis, high conductivity, low density, and high specific surface area. The suitability of various carbon-based architectures for sulfur cathodes has been investigated [23–26].

They can be categorized into a number of forms according to their physical structure, as shown in figure 4.3.

The major role of a carbon host in a sulfur cathode is:

(i) Physical confinement as well as chemical anchoring of sulfur species;
(ii) Providing a mechanical and conductive scaffold;
(iii) Providing a more electrochemically accessible active surface area for functioning of LiSBs.

However, selection of a carbon host can be made based on the following criteria:

- High specific surface area for electrochemical activity;
- Strong interaction between sulfur and carbon for chemical anchoring of sulfur species;
- Porous structure with desirable pore size, shape and distribution.

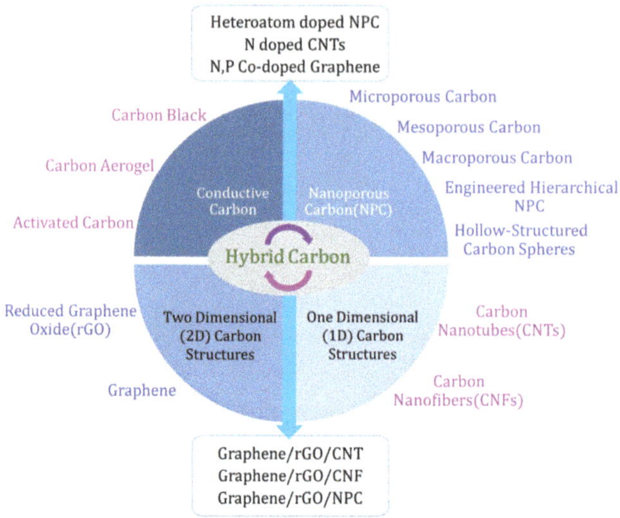

Figure 4.3. Schematic illustration of various carbon materials as sulfur hosts in LiSBs.

4.5.1 Nanoporous carbon

Carbon materials containing pores and voids in their structure are treated as nanoporous carbon materials. Depending upon pore size they can be classified into three major sets: microporous, mesoporous, and macroporous with pore sizes less than 2 nm, 2 to 50 nm, and greater than 50 nm, respectively.

The synthesis route and source for carbon have played a significant role for sulfur stacking into the pores. Most of the porous carbon materials have been prepared via conventional routes such as chemical and physical activation, catalytic activation by organometallic compounds, using biomass, carbonization of polymers, etc. Nowadays, some template-assisted methods have been reported to obtain engineered hierarchical porous carbon materials. This approach helped significantly to manage pores with desirable shape, size and their distribution. In general, zeolites, metal–organic frameworks (MOFs), and ordered mesoporous silica as template, are the most preferable representatives for template-assisted synthesis [9, 12, 24–26].

Microporous carbon

Microporous carbon can offer strong spatial confinement of sulfur species and effective loading of elemental sulfur, due to comparable pore size (less than 2 nm) and huge specific surface area about 1000 $m^2\ g^{-1}$ [27–30].

It has been predicted that a pore with a diameter in the range of 0.5–0.7 nm can trap the low-order polysulfides. Thus, an S@microporous carbon composite can help to overcome the problem of the ‘shuttling effect’ due to the dissolution of polysulfides, leading to enhancement in specific capacity and cycle stability [31].

Hu *et al* reported a series of microporous carbons synthesized via immediate carbonization and self-activation strategy of alkali-complexes with good rate capability and cycle life. The reversible specific capacity of 968.4 mAh g^{-1} at 0.1 C and 616.3 mAh g^{-1} at 1 C were obtained after 50 and 600 cycles, respectively. The excellent performance was attributed to the substantial confinement of low-ordered PSs in an ultra-microporous structure (size < 0.7 nm) [31].

Elazari *et al* prepared microporous carbon fibers with a specific surface area of about 2000 m^2 g^{-1} and pore volume of 0.57 cm^3 g^{-1}. The cathode showed a specific capacity of 800 mAh g^{-1} at 0.09 C over 80 cycles with sulfur loading of about ~33.3%. Even with low sulfur loading, the excellent performance obtained was attributed to a more active electrode/electrolyte interface due to good wettability and excellent charge kinetics through fibers [32].

Zhu *et al* reported the synthesis of ultra-microporous carbon (UMC) structures by the pyrolysis method using polyvinylidene fluoride. A UMC with uniform pore size of about 0.55 nm was reported for effective encapsulation of low-ordered PSs (S_2–S_4). The cathode exhibited an excellent specific discharge capacity of 852 mAh g^{-1} at 0.1 C and coulombic efficiency of about 100% after 150 cycles. It also showed decent cyclic stability with 0.03% capacity decay per cycle at 1 C for 1000 cycles. The excellent performance of the sulfur/UMC nanocomposite cathode was attributed to efficient trapping of sulfur species and diminished shuttling effect [33].

Moreover, a discharge curve of a S_2–S_4/microporous carbon cathode showed only one plateau at ~1.85 V instead of two at 2.4 and ~2.2 V for the conventional sulfur cathodes. This means that there may be a direct reduction of S_2–S_4 PSs to Li_2S without electrolyte soluble intermediate LPSs, which was advantageous to improve the electrochemical performance of the LiSBs [34].

However, there are certain issues associated with microporous carbon too, such as [35–41]:

- Smaller pore size causes less sulfur loading (less than 40%), resulting in low capacity and energy density.
- The higher-order LPSs have a size greater than that of elemental S_8, and thus may not be accommodated in the micropores, causing instability of the S cathode. Very small micropores have less accessibility for the electrolyte causing a decrease in sulfur utilization.
- Low mass transport of Li ions, resulting in poor rate capability.

Mesoporous carbon

The problem of low sulfur content in microporous carbon can be overcome with mesoporous carbon, which offers a high surface area and high pore volume, with pore sizes between 2–50 nm. Several mesoporous carbon materials have been examined as host materials for S cathodes in LiSBs with high sulfur content of about 60 to 85% and superior electrochemical performance with a high reversible specific capacity of 1005 mAh g^{-1}. The mesoporous confined S cathode is readily available during operation of LiSBs, due to good contact between the electrolyte and active electrode material. So, a mesoporous carbon host is expected to demonstrate high energy density and good cycling performance because of good structural and electrochemical stability, high sulfur content, high electrical conductivity, and the capability to capture dissolved LPSs [42, 43].

In the early days, Nazar's group reported a CMK-3/sulfur mesoporous carbon framework by a melt-diffusion method with sulfur content of about 70 wt% in the electrode. The composite also exhibited high electrical conductivity, high specific surface area, and desirable pore volume with values of about 0.2 S cm^{-1}, 1976 m^2 g^{-1}, and 2.1 cm^3 g^{-1}, respectively. The cathode showed a high reversible capacity of 1100 mAh g^{-1} at 0.1 C after 20 cycles [44].

Li *et al* prepared a mesoporous carbon series by a two-step impregnation process as a sulfur host in LiSBs. A series of composites were prepared with pore sizes of 7, 12, and 22 nm, pore volumes of 1.72, 3.64, and 4.80 cm^3 g^{-1}, and high sulfur content of about 66, 80, and 83 wt%, which exhibited an initial capacity of 1195, 1071, and 1050 mAh g^{-1}. The lesser capacity even with high sulfur content was attributed to an increase in internal resistance. The same host possessing a pore size of 22 nm if filled with low sulfur of about 50% showed an increased initial capacity of 1390 mAh g^{-1} and reversible capacity of 840 mAh g^{-1} after 100 cycles. The enhanced performance was attributed to partial filling of mesopores, which further leads to high sulfur utilization and efficient accommodation of high-order LPSs [45].

Three-dimensional mesoporous structures also showed promising results. For instance, Wang *et al* fabricated a sulfur cathode with sulfur/mesoporous carbon spheres with a hierarchical porous structure, which exhibited a specific capacity of ~1390 mAh g^{-1} due to high conductivity and fast ion diffusion process [46].

The template-assisted synthesis route has also been found to be effective for the preparation of ordered mesoporous carbon materials. Carter *et al* synthesized mesoporous carbon using a porous silica template. The mesoporous carbon/sulfur composite cathode exhibited high electrochemical performance with initial discharge specific capacity of about 1350 mAh g^{-1} at 0.1 C and good retention with reversible capacity of about 830 mAh g^{-1} at 0.2 C after 100 cycles. The enhanced performance was attributed to the strong interaction of LPSs with the oxidized carbon surface, higher conductivity, improved ion diffusion, and faster electron transport through well-defined ordered mesoporous carbon [47]. Significant electrochemical properties of sulfur cathode with mesoporous carbon/sulfur nanocomposite through high sulfur content and with different carbon structures like mesoporous carbon spheres, mesocellular carbon foam, mesoporous carbon nanosheet-like honeycomb structure, 3D mesoporous carbon frameworks, etc, have been reported and their performance is summarized in table 4.1 [47–53]. Some studies on composites of mesoporous carbon/graphene and micro/mesoporous carbon spheres showed an initial capacity in the range of 1500 to 800 mAh g^{-1}.

Thus, in summary, the noteworthy enrichment in electrochemical performance in terms of excellent rate ability and long life cycling stability with a mesoporous carbon host is attributed to:

- Hierarchical porous structure for the confinement of high sulfur;
- Easy accommodation of higher-order LPSs in desirable pores with larger pore size;
- Higher conductivity of well-defined ordered mesoporous carbon;
- Improved ion diffusion and faster electron transport through the porous network;
- Efficient utilization of active material.

However, mesoporous carbon with uniform pore structure, correlative pore volume, and surface area is hardly achievable. Also, the presence of micropores is inevitable, also their density cannot be monitored or controlled.

Macroporous carbon
Macroporous carbon materials have a pore size greater than 50 nm. The advantage of macroporous carbon compared to micro and mesoporous carbon materials is that it holds more sulfur due to high pore volume. Additionally, the specific capacity of a sulfur cathode with a macroporous host is decided mainly by the mass of the S@C composite. However, larger pore size resulted in poor confinement of polysulfide, leading to poor cyclic stability with fast fading rate [54, 55].

In summary, nanoporous carbon with micropores has less sulfur loading (less than 40%), and also does not confine higher-order LPSs, but has strong interaction with low-order polysulfides. Mesopores offer higher conductivity of well-defined ordered mesoporous carbon but the presence of micropores in it is inevitable. Macropores hold more sulfur due to high pore volume but have poor confinement of polysulfide. Thus, the ordered hierarchically micro-macro-/mesoporous architecture can scaffold the enhanced performance of LiSBs, due to the synergistic effect of high sulfur loading and significant trapping of sulfur species. Table 4.1 provides an overview of the electrochemical performance of sulfur/porous carbon nanocomposite cathodes.

4.5.2 Engineered hierarchical nanoporous carbon

Considering the merits and limitations of various pore sizes, it is expected that the hierarchical nanoporous carbon host materials with optimal micro/meso/macropores can demonstrate superior electrochemical properties. Due to moderate pore volume they can easily accommodate the low as well as high-order polysulfides, reduce the shuttling effect, fast ion diffusion and electron transport through the conducting porous network, efficiently utilize active material and withstand volume expansion. According to earlier findings, aqueous electrolytes can benefit from pore sizes of 0.4 or 0.7 nm, while organic electrolytes may require pore sizes of 0.8 nm. The ideal choice of carbon electrode for LiSBs will therefore have a hierarchical pore structure with micropores for charge storage, mesopores for ion transport, and macropores for electrolyte penetration. Several hierarchical porous carbon materials with micro/meso/macropores have been synthesized to improve the performance of LiSBs. All of these approaches mainly focus on significant encapsulation of sulfur species, enhancement of specific surface area, optimization of pore size, and optimized sulfur loading.

For instance, Jung *et al* [56] prepared a sulfur/hierarchical porous carbon nanocomposite. The composite showed a remarkable improvement in cyclic performance. The study reported desirable retention stability over 500 cycles at 2.4 C with a discharge specific capacity of roughly 540 mAh g^{-1}.

Li *et al* [38] synthesized a sulfur/ordered meso/microporous carbon nanocomposite electrode, which demonstrated a specific capacity of 837 mAh g^{-1} and retention

Table 4.1. Performance of LiSBs using various nanoporous carbon host materials for the sulfur cathode.

Cathode/sulfur host	Initial specific discharge capacity mAh g^{-1}@C	Reversible capacity mAh g^{-1}@C	Cycles // capacity retention (%)@C	Sulfur content [wt%]	Sulfur loading [mg cm^{-2}]	References
Microporous C	616.3@1		600			[31]
Microporous C fibers	800@0.09		80//33.3			[32]
Ultra-microporous C (UMC)	852@0.1		150//100			[33]
S@C hybrid	1040@0.1	700	500	70		[34]
Porous C	1382@0.5		1000//61		2.4	[35]
C@S		974@0.5	100	70		[36]
Ultra-microporous C (UMC)	1000@1.0	510	100	32	1.0	[33]
Macroporous CCo@N	1150@1		300//57			[37]
Ordered meso@microporous C	1037@0.5	837	200	60.6		[38]
N-doped microporous C	1382@0.3		200//72	53	1.2–1.4	[39]
Micro/mesoporous C fiber	1071@0.5		100//88		2.0	[40]
Porous CNSs@G aerogels	1328@0.2		300//49	57	2.0	[41]
Mesoporous C series	1390	840	100	50		[45]
S/Mesoporous C spheres	1390					[46]
Oxidized mesoporous C derived porous silicon	1350@0.1		100// 81@0.2	60	0.3–0.6	[47]
N-doped mesoporous C	1585@0.1		100//54			[48]
N&O dual-doped nonporous carbonaceous material	654@1.6 A g^{-1}	540	300	52.5	1.39	[49]
N&O dual-doped porous C	1185@0.2		600 // 62@1.0		12–1.5	[50]
Co-embedded N, P-doped porous C	480@5.0	411	400	56	0.8–1.0	[51]
N-doped porous C nanowire	1430@0.2				1.5–1.7	[52]
HPC Encapsulating S	1249@0.1	831	100	84		[53]
S@HPC nanocomposite	539		500//77 @2.4			[56]
S@Meso@Microporous C nanocomposite	837		200 // 81@0.5			[38]
Macro@Microporous C	1241	791.6	100@0.5			[58]
Micro@Meso@Macroporous C material using $CaCo_3$	1327	630	100@0.2			[60]
3D flower-shaped activated porous C@S	1388@0.1	800	50	60		[64]
3D vertically-aligned interconnected porous C	1208	837	300	68		[65]
3D porous C Composite	1242@1	927	200	90		[66]
S@C sphere		650@0.23	500	42		[67]

(*Continued*)

Table 4.1. (*Continued*)

Cathode/sulfur host	Initial specific discharge capacity mAh g^{-1}@C	Reversible capacity mAh g^{-1}@C	Cycles // capacity retention (%)@C	Sulfur content [wt%]	Sulfur loading [mg cm^{-2}]	References
C microspheres	728@2.0	449	2000	48	0.3–0.5	[68]
Hollow core–shell sphere	1140@0.5	960	200	70		[69]
Porous hollow indented C sphere	1478@0.1		1200 // 46@0.5	75		[70]
B-doped porous C sphere@G	740@0.5	555	500	49	1.0–1.3	[71]
N&P co-doped C sphere (NPDSCS)	952@1.0	814	500	58	5.8	[72]
HPC aerogels	1279@0.2		100 // 78		1.3–1.6	[73]
3D interconnected porous C aerogel	1441@0.05	820	50	35		[74]
C microgels	1132@0.1		200 // 74			[75]
Sheet-like C matrix	1099@0.2		100 // 55		2.0	[76]
N-doped C foam	1124@0.5		100 // 80		1.2	[77]
CNT@CNF@MnO_2 nanosheets	780@2.0		300 // 64		2.4	[78]
3D porous C_3N_4 nanosheets @rGO (PCN@rGO)	1025@0.1	411	200	55	4.0	[79]
rGO@G-C_3N_4@CNT	1017@0.2	540	300	56.6	4.2	[80]
NG@G-C_3N_4 hybrid C nitride	1100@0.5		100	65.5	2.0	[81]

of about 81% at 0.5 C over 200 cycles. The good cyclic stability in both cases was attributed to the hierarchical pore network, optimal density of mesopores/micropores, high specific surface area, and large pore volume.

Liang *et al* [57] reported a high initial specific capacity about 1585 mAh g^{-1} with a sulfur cathode with hierarchical nanoporous carbons and homogeneous meso/micropores. The increased capacity was attributed to selective filling of the micropores and strong interaction between sulfur and the carbon host.

Yu *et al* [58] prepared a macro/microporous carbon framework via a one-step approach of carbonization/activation. The activation process caused enhancement in the specific surface area of about 1211.6 m^2 g^{-1} with pore volume about 0.59 cm^3 g^{-1}, which resulted in a high initial specific capacity of 1241 mAh g^{-1} and discharge capacity of 791.6 mAh g^{-1} over 100 cycles at 0.5 C. The enhanced performance was also attributed to strong chemical anchoring of elemental sulfur on the oxygen-functionalized carbon structure. It has been reported that the optimization of chlorination temperature between 700 and 900 °C leads to precise control of size distribution of the pores between 0.7 and 4.0 nm, which results in a reduction in defects in the carbon structure and sufficient sulfur loading [59].

Tang *et al* synthesized micro/meso/macroporous carbon material using $CaCO_3$ as a hard template. The sample, prepared at a carbonization temperature of about 950 °C, exhibited initial specific capacity of 1327 mAh g^{-1} and reversible capacity of 630 mAh g^{-1} after 100 cycles at 0.2 C. The authors observed that, the value of specific capacity for electrodes with the same sulfur content and pore size distribution has substantially more influence than the total pore volume and specific surface area [60].

Hollow nanoporous carbon core–shell structures with macroporous core and micro- or mesoporous shell have been used in several investigations as sulfur host, exhibiting good cycle stability and a high specific capacity. In the core–shell structure, macroporous 'core' serve well for sulfur loading and also provide structural stability during volume expansion, while the micro- or mesoporous shell can provide trapping centers which act as a physical barrier to prevent the diffusion of polysulfides, and effective electrolyte penetration through the continuous porous network leads to fast ion and electron transport kinetics [59, 61]. Several studies reported hollow carbon spheres with a core diameter of around 200–500 nm and a mesoporous shell with a thickness of 30–50 nm. A thin shell ensures quick Li-ion diffusion and electron-transfer reactions, while large core volume promotes sulfur loading and accommodates long-chain PSs [62, 63].

However, since all carbon structures are nonpolar in their pure state, they interact poorly with polar sulfur molecules. Physical confinement, which may rupture during numerous charge–discharge cycles, is the primary way of the majority of sulfur anchoring in pure carbon.

According to section 4.3.2, strong confinement and better cyclic behavior can be achieved by the polar–polar contact between the host and sulfur species. Carbon materials containing either non-polar functional groups or appropriate heteroatoms can be effectively converted to polar. Chemical anchoring of sulfur species through polar–polar interaction with polar functional groups leads to improved cyclic stability and rate performance. Strong polar–polar interaction with PSs is made possible by the polar functional groups via chemical bonds that are formed when sulfur is chemically bound to carbon and oxygen (i.e. C–S and S–O). Similarly, heteroatom-doped carbon structures have been investigated with cations such as nitrogen, phosphorus, boron, oxygen, etc [48–52]. Among these, the nitrogen-doped hollow nanoporous carbon structure showed a promising performance for Li–S batteries in terms of specific capacity and conductivity. The instilled polar nature of carbon materials was attributed to the formation of different chemical environments by nitrogen like 'pyridinic N', 'graphitic N', and 'pyrrolic N' [49, 50, 52]. Table 4.1 summarizes the performance of LiSBs using various nanoporous carbon host materials including engineered hierarchical nanoporous carbon host materials based on numerous investigations [53–81].

4.5.3 One-dimensional CNT and CNF based cathodes

One-dimensional carbon materials have been used as sulfur hosts and conductive additives in the cathode to improve electrical conductivity and reaction kinetics in LiSBs. They mainly offer a continuous 1D channel for efficient charge transfer, good mechanical support to the cathode, and tunable properties with respect to their structural parameters. Due to their hollow structure they can accommodate sulfur species on the surface as well as in the hollow inner space of their structure. The most commonally studied 1D carbon structures for sulfur cathode are carbon nanotubes (CNTs) and carbon nanofibers (CNFs) [82, 83].

The fullerenes family of carbon have hollow structures with different shapes like a hollow sphere, ellipsoid, or tube. Tubular fullerenes are known as CNTs because of their tube-shaped geometry with diameters in the nanometric range. Since their discovery in 1991 by Japanese electron microscopist *Sumio Iijima*, CNTs have attracted the attention of researchers because of their extraordinary properties such as [84–86]:

- High aspect ratio with 132 000 000:1 (length-to-diameter ratio);
- High electrical conductivity;
- Good mechanical strength and feasibility;
- Ability to act as thermal conductors as well as thermal insulators;
- Tunable structural properties (dependent of radius and cylinder rolling angle);
- Versatile electric properties (armchair CNT: metallic; chiral CNT: semiconducting);
- Ability to carry electrical current density of 4×10^9 A cm^{-2};
- High thermal stability up to 750 °C, which enables a wide range of temperature performance and safety of LiSBs;
- Effective physisorption and chemisorption of chemical as well as biological species;
- Light weight with high surface-to-volume ratio;
- Ability to relieve internal polarization.

CNTs are mainly observed in the form of sp^2 hybridization with the hexagonal lattice of side-walls. They can exist in two structural forms, single-wall carbon nanotubes (SWCNTs) and multi-wall carbon nanotubes (MWCNTs) [85, 86].

It is expected that the high length-to-diameter ratio can be beneficial for free-standing CNT based cathode fabrication. The S@CNT composite cathode can be prepared using two approaches. In the first approach, the CNT porous framework is produced, and then elemental sulfur is introduced into it either using melting-diffusion or chemical deposition techniques. While in the second, the CNTs and elemental sulfur are embedded in a specific porous host matrix. The first one is found to be very successful for sulfur loading and charge transport kinetics [87–89]. To investigate the usage of CNTs in various composites as sulfur hosts or conductive additives in sulfur cathodes, several studies have been conducted. With a S@CNT composite cathode, a large initial capacity of 1600 mAh g^{-1} has been recorded, with only one plateau at roughly 1.85 V. The increased loading of sulfur (approximately 95%), effective trapping of low-order PSs, and shorter path lengths offered by the continuous 3D conducting network of CNTs for effective ion diffusion and charge transport were all credited with the increased capacity [85, 88, 90, 91].

Additionally, it has been noted that CNTs with small apertures are advantageous for the effective containment of low-order sulfur species and their subsequent use during redox processes [91, 92]. However sulfur adhered on the outer wall of a CNT does not effectively avoid the dissolution of the PSs, resulting in rapid capacity fading [93, 94].

Through the use of CVDs and template-assisted techniques, highly pure CNTs with good aspect ratio have been synthesized by several groups. Additionally, CVD-prepared CNTs are being marketed for sale. With high sulfur loadings of around 70%, CNTs produced directly on a substrate by CVD demonstrated good electrochemical activity. In addition to having a 3D-ordered vertical porous network that is advantageous for quick electron transmission and appropriate sulfur loading, the aligned CNTs also have good mechanical stability due to their proper contact with the substrate [91–94].

For instance, Cheng *et al* prepared vertically-aligned CNTs directly on a metal substrate by CVD. The cathode exhibited sulfur loading of about 90%, which resulted in high volumetric capacity of about 1116 mAh cm^{-3} and energy density of 1249.2 Wh kg^{-1} at a power density of 134.3 Wh kg^{-1}. The enhanced performance was attributed to good electrical conductivity of CNTs and successive conductive network for fast ion diffusion [95]. However, the relatively low coulombic efficiency was due to shuttling of LPSs through the open porous network of CNTs.

Su *et al* reported a self-weaving S-MWCNT nanocomposite as a cathode with high specific capacities of 1352 mAh g^{-1} at 1 C and 1012 mAh g^{-1} at 4 C. The MWCNTs offer an enormous specific surface area, which confines the PSs and prevents the loss of the active ingredients, which was attributed to the enhanced performance [96].

Yuan *et al* compared the performance of S@CB and S@MWCNT composite electrodes. The S@MWCNT composite exhibited a reversible capacity of about 670 mAh g^{-1} after 60 cycles; this is around three times greater than that of the S@CB composite electrode,which had a value of 270 mAh g^{-1} after 40 cycles at 100 mA g^{-1}. The enhanced performance with S@MWCNT composite electrodes was attributed to fast ion diffusion and charge transport through the short and conducting pathway facilitated by the MWCNT network [97].

Chen *et al* fabricated a core–shell structured S@MWCNTs-microporous carbon composite architecture as the S cathode with a sulfur loading of roughly 57%. The MWCNTs had 3% functionalization of carboxylic groups, which aided in the efficient chemical anchoring of the sulfur species. The composite electrode exhibited reversible capacities of about ~1000 mAh g^{-1} and 780 mAh g^{-1} after 100 cycles at ~0.18 C and 200 cycles at 0.3 C, respectively. The short Li-ion diffusion pathway and significant sulfur species encapsulation in MWCNT, which minimized the shuttling effect, were responsible for the improved excellent cyclic performance [98].

The S cathode in LiSBs performed electrochemically better as a result of heteroatom doping in CNT. Heteroatom-containing precursors, such as those that contain N, O, P, and B can be heated to perform the doping. As an illustration, nitrogen doping in CNTs also showed a much better performance due to higher electrical conductivity, strong confinement of PSs, and high electrolyte penetration as a result of improved wettability [99–101]. A S@hydroxylated MWCNT composite also demonstrated increased cycle stability and rate capability because of a decreased shuttling effect caused by the significant chemical interaction of LPSs with the hydroxyl functionalized MWCNT surface [89, 102].

Additionally, the B and O dual heteroatom doping in MWCNT demonstrated a synergistic effect of improved charge transport capability and strong chemisorption to the surface of CNTs, leading to high electrochemical performance with excellent reversible capacity in the range of 950 to 800 mAh g^{-1} [89].

Despite the improved electrochemical characteristics of the S@CNT nanocomposite cathode, the problem of the shuttling effect persists because of the weak space confinement of high-order LPSs, which has a negative impact on cycling stability.

Another highly explored 1D carbon structure for sulfur hosting is sp^2 and sp^3 hybridized carbon nanofibers (CNFs) because of their great ability to conduct electrons due to their fibrous structure, high specific surface area, and 3D highly accessible porous network without a hollow region in the center. Additionally, CNFs offer larger pore volume within interlocked sheets that is helpful for absorbing polysulfides as well as a continuous and shorter channel for ion diffusion and electron transport [89, 103, 104]. Electrospinning, partial decomposition of the PMMA blend, carbonization of polyacrylonitrile (PAN) and polymer nanofibers, the template-assisted method employing a silica template, and the chemical solution reaction method are the main methods used to synthesize nanoporous CNFs.

The S@CNF nanocomposite cathode with a sulfur loading of roughly 42% fabricated using a low-cost electrospinning method has a specific surface area of 1928 m^2 g^{-1}, a large pore volume of 2.41 cm^3 g^{-1}, and exhibited a discharge capacity of 1400 mAh g^{-1} at 0.05 C [105]. Similar to CNTs, the structures of CNFs can be rehabilitated by surface activation, coating, and by heteroatom-doping to improve their electron affinity for strong confinement of sulfur species and their wettability for easy electrolyte penetration, which leads to a high initial discharge capacity of about 1500 mAh g^{-1} [106–111]. Table 4.2 summarizes the electrochemical performance of the CNT, CNF, and their nanocomposite as sulfur hosts [95–121].

In conclusion, the following properties of one-dimensional CNTs and CNFs make them potential sulfur host materials:

- Excellent structural qualities with more electro-active sites benefit from a high aspect ratio;
- Benefits of an interconnected 3D porous architecture include increased sulfur loading and the confinement of short-chain PSs;
- High conductivity for increased energy density and exceptional capacity;

Table 4.2. Electrochemical performance of the CNT, CNF, and their nanocomposite as sulfur hosts.

Cathode/sulfur	Initial specific discharge capacity mAh g^{-1}@C	Reversible capacity mAh g^{-1}	Cycles // capacity retention(%)@C	Sulfur content [wt%]	Sulfur loading [mg cm^{-2}]	References
S@MWCNT Microporous C		1000	100@0.18		57	[98]
N-doped porous C	957@0.05		100	80		[99]
P-doped CNT	1106@0.1		100//75		1.3	[100]
CNT@Micro-mesoporous C	1670@0.1	1142	200	32	1.0	[112]
MWCNT porous microspheres	983@0.2		100//87		2.5	[113]
Hierarchical CNT@S paper electrodes	1023@0.1	700	200		1.0	[87]
S@Mesoporous CNT	1109@0.1	760	100	70		[88]
CNT@C nanocage	1354@1.0	892	50	60		[114]
CNT–G@S	911.5@0.2	771.8	100	53		[115]
CNFs	753@0.1	680	100	79	10.5	[116]
Silica@S@C fiber	943@0.5		100//51		1.3	[117]
Multichannel CNF (MCNF)	990@0.5	753	200	64	4.6	[118]
1-D HPC Fibers	535@15		500//62@5			[119]
Hollow CNF@S	1090@0.1	766	100	60.8		[120]
N-doped porous CNFs	1564@1.0	562	50	60		[121]
S@G	1265.3@0.1	679.7	500	70	2.0	[142]
rGO@G foam	1000@0.2	645	350	83	9.8	[143]
Hierarchical 3D N@P@G@CNT	826.1		48.29	70		[141]
S@2D-GNS	1611	580	40			[130]
S@Porous G	1500	4.2 mAh/cm	400	80		[131]
N-doped G-G@S	1200@0.3 A g^{-1}	700	300	70	2.0	[144]
	1430@0.5 A g^{-1}			65		[145]

(*Continued*)

Table 4.2. (*Continued*)

Cathode/sulfur	Initial specific discharge capacity mAh g^{-1}@C	Reversible capacity mAh g^{-1}	Cycles // capacity retention(%)@C	Sulfur content [wt%]	Sulfur loading [mg cm^{-2}]	References
G-directed 2D porous C frameworks						
rGO nanoscroll	1295@0.2		100//57	76	0.8	[146]
N-doped G	1082@0.1		200	90		[135]
Hollow CNFs@rGO	1318@0.2				1.4	[147]
rGO@S	960@0.1		500	90		[148]
rGO@S	1340@0.1		100//52		1.02	[149]
Porous CNSs@G	1328@0.2		300//49	57	2.0	[150]
G@Microporous C	1300@0.1			75	1.5	[151]
GO@CNT	650@0.1	500	500	70	4.0	[152]
G Nanosheets (GNS)	758@0.5	450	500	53.5	1.3–1.6	[153]
N-doped CNT@G	835@1.0	635	400	49.1	4–1.8	[154]
N-rich CNT@G	1314@0.2		200//97@2	73	1.5	[155]
3D G@CNT@S hybrid structure	1008@0.3	704	100	70.3		[156]
CNT@GNS	829@0.5	716	500	81		[157]
Super-aligned CNT@G hybrid	1048@1.0	594	1000	50		[158]
Super-long CNT@G spheres	829@0.5	640	500	64		[157]
G@mesoporous CNS	900@1.0	719	500	73	5.0	[159]
B-doped porous C sphere@G	740@0.5	555	500	49	1.0–1.3	[71]
G sponge@S	625		300 //70	80	12	[160]
N-doped G foam	1219		300		8.2	[161]
rGO aerogel	532@0.1		300//85		6.4	[162]

- Easy synthesis directly over the substrate decreases the contact resistance and facilitates the shortest channel for quick charge transport and ion diffusion;
- Large surface area and pore volume for uniform distribution and dispersion of sulfur species.

4.5.4 Graphene-based materials as sulfur hosts

Because of the discovery and development of adaptable classes of various materials, the 21st century might be considered the age of innovative functional materials. Graphene, one of the most intriguing materials among them, has the potential to bring about an incredible breakthrough in science and technology. It is a hexagonal lattice made up of a sheet of carbon monolayer [122–124].

It has been regarded as a promising sulfur host material for the S cathode due to its distinctive physical characteristics, such as: vital 2D structure with large theoretical specific surface area (2600 $m^2 g^{-1}$), high charge carrier mobility (15 000 $cm^2 V^{-1} s^{-1}$), high electrical conductivity (10^4–10^5 S m^{-1}), good thermal conductivity (5×10^3 W $m^{-1} K^{-1}$), excellent flexibility and good mechanical strength (1 TPa) [125–129].

Graphene with various graphitic structures, including films, spheres, foams, and sponges, has been reported as a sulfur host. These graphene materials were synthesized using a variety of techniques, including freeze-drying, epitaxial growth, vacuum filtration, and metal reduction self-assembled methods. Due to their large specific surface area and effective pore volume, the 3D conductive skeletons of graphene were found to be more efficient than stacked graphene in terms of encapsulating sulfur nanoparticles, transporting ions and electrons, and preventing the dissolution of polysulfides during charge–discharge. Additionally, it has been reported that the degree of graphitization considerably affects the electrical conductivity of graphene [124–129].

Wang *et al* reported a melt-diffusion-produced S@2D graphene nanosheet (GNS) composite with an initial discharge capacity of about 1611 mAh g^{-1} at 50 mA g^{-1} and a discharge capacity of approximately 580 mAh g^{-1} after 40 cycles. The performance was attributed to enhanced electrical conductivity and high specific surface area beneficial for trapping LPSs [130].

Deng *et al* prepared a sulfur cathode based on a S@porous graphene composite with high pore volume up to 3.51 $cm^3 g^{-1}$ and sulfur content around 80 wt%, which exhibited an initial discharge specific capacity of 1500 mAh g^{-1} and a better areal capacity of 4.2 mAh cm^{-2} after 400 cycles [131].

The cyclic performance was attributed to the highly conductive hierarchical graphene structure which contributed to fast kinetic processes. Fang *et al* synthesized a series of free-standing 3D graphene frameworks (3DGFs) as sulfur hosts by a facile one-pot method. Among the prepared samples, 3DGF-S90 with 90% sulfur loading exhibited a specific capacity of 969 mAh g^{-1} at 0.1 C and good capacitance retention with a fading rate of about 0.052% per cycle over 500 cycles. The three-dimensional porous framework facilitated high sulfur loading of 80–90% and also provided an effective channelized pathway for the rapid ion diffusion process. The free-standing 3DGF with high sulfur also inhabited the sulfur species, thus reducing the shuttling effect [132].

Zhou *et al* prepared a S@fibrous graphene foam composite for the S cathode by a facile one-pot hydrothermal reduction method, which demonstrated high capacity of 1160 mAh g^{-1} at 0.3 A g^{-1}; however, limited confinement of LPSs caused rapid capacity fading [133].

Similar to other carbon-based materials, graphene can be functionalized by heteroatom-doping to improve the chemical anchoring of the sulfur species [134, 135]. For instance, Qiu *et al* demonstrated enhanced initial discharge capacity with a nitrogen-doped graphene sulfur host of around 1200 mAh g^{-1}. With a discharge capacity of 760 mAh g^{-1} over 120 cycles at 0.2 C, the composite also demonstrated good cycling performance. The enhanced performance was attributed to improved adsorption sites for polysulfides via strong chemical interactions [136].

However, an inherent self-restacking issue due to a strong π–π interaction and Van der Waals forces between succeeding graphene layers results in a reduction in specific surface area, which has an impact on charge transport and the kinetics of ion diffusion. Additionally, it is still difficult to produce graphene in large quantities at a competitive price with repeatable features, which restricts its practical use. Graphene oxide (GO) and reduced graphene oxide (rGO), as modified version of graphene, have opened new opportunities for the sulfur host in LiSBs. GO can be bulk manufactured by thermally exfoliating graphite with a strong oxidizing agent. Because both rGO and GO have several functional groups, including epoxide, carbonyl, hydroxyl, and phenol on their surfaces, it has been observed that they are effective for the confinement of sulfur species [129–137].

The cathode of the GO-wrapped S nanocomposite performed better due to effective S hosting and trapping of soluble LPSs. Through chemical confinement of soluble LPSs via oxygenated functional groups, the unstacked GO suppresses the shuttle action. Additionally, accommodating the volumetric strain, the wrapped GO gives the cathode structural stability. For instance, Ji *et al* reported an improved performance of a S@GO composite cathode with a high reversible capacity of about 1000 mAh g^{-1} after 60 cycles and capacity retention of 94.5% over 1500 cycles with a fading rate of about 0.039% per cycle [138]. The improved performance was attributed to reduced LPS dissolution due to chemical anchoring of the sulfur species with oxygenated functional groups. Cation doping or the use of cationic surfactants have been used for surface functionalization of the GO surface. Heteroatoms were added to the graphene structure, which resulted in significantly improved electrical and electrochemical characteristics and superior chemical anchoring of the sulfur species [136, 139, 140].

Due to a synergistic effect, various porous carbon–sulfur–GO nanocomposites can exhibit excellent electrochemical performance as the cathode material, overcoming the limitation of individual pristine carbon structures. Numerous attempts have been made to use various combinations of carbon materials as a host for sulfur, including graphene/CNTs, graphene/CNFs, graphene/nanoporous carbon, CNT/CNF, GO/porous carbon, CNT/mesoporous carbon, and so on.

For instance, Walle *et al* [141] prepared hierarchical three-dimensional nitrogen and phosphorous co-doped graphene and carbon nanotubes with 70 wt% of sulfur content (N, P co-doped G/CNTs-S70) composite using melamine phosphate (MP) as a single precursor of N and P. An X-ray photoelectron spectroscopy (XPS) study was carried out to analyze the elemental composition in a N, P co-doped G/CNTs-S70 composite, through critical investigation. Figure 4.4 displays the N 1s and P 2p spectra and MP (source of nitrogen and phosphorous) of the N, P co-doped G/CNTs-S70. The N 1s spectra of the N, P co-doped G/CNTs-S70 in figure 4.4(a) show three peaks at 399.6, 400.0, and 401.7 eV, which correspond to pyridinic N, pyrrolic N, and graphitic N, respectively. These functional groups serve as anchoring sites for chemical adsorption of PSs. In contrast to the N 1s spectrum of MP (figure 4.4(c)), pyridinic N's binding energy shifts to a lower energy, quaternary N's binding energy shifts to a higher energy, and a new peak of pyrrolic N forms, signifying the existence of a robust chemical interaction between N and the carbon scaffold matrixes. The XPS spectra of the P 2p of N, P co-doped G/CNTs-S70 are shown in figure 4.4(b) and exhibit two peaks at 131.5 and 133.1 eV, respectively, that correspond to P–C and P–O bonds. The hybridization between P 3p and C 2p orbitals might be formed by converting sp^2 carbon into sp^3 hybridization. Compared with the P 2p spectrum of MP (figure 4.4(d)), the binding energy of P–O shifts to a lower energy and a new peak of P–C is formed, indicating a strong interaction between the P atom and carbon scaffold matrixes.

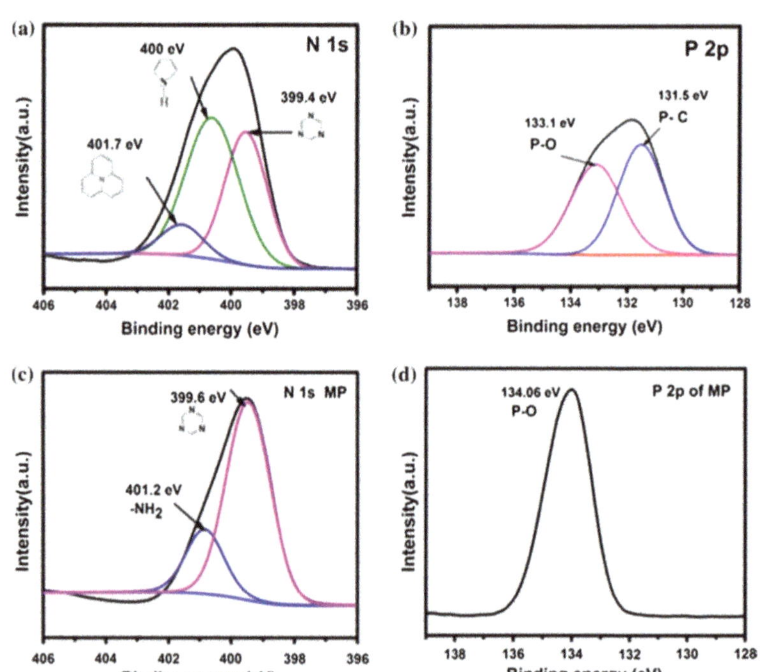

Figure 4.4. High-resolution XPS spectra of (a) N 1s peaks of N, P co-doped G/CNTs-S70, (b) P 2p peaks of N, P co-doped G/CNTs-S70, (c) N 1s peaks, and (d) P 2p peaks of melamine phosphate (MP). Reprinted from [141], copyright (2017) Springer Science Business Media, LLC, with permission from Springer Nature.

Figure 4.5 displays the electrochemical performance of both samples, as determined by the GCD charge–discharge profile. The cycling performance of N, P co-doped G/CNTs-S70 and G/CNTs-S70 composite electrodes at a discharge rate of 0.5 C is shown in figure 4.5(a). A high initial discharge capacity of 1550 mAh g^{-1} and a

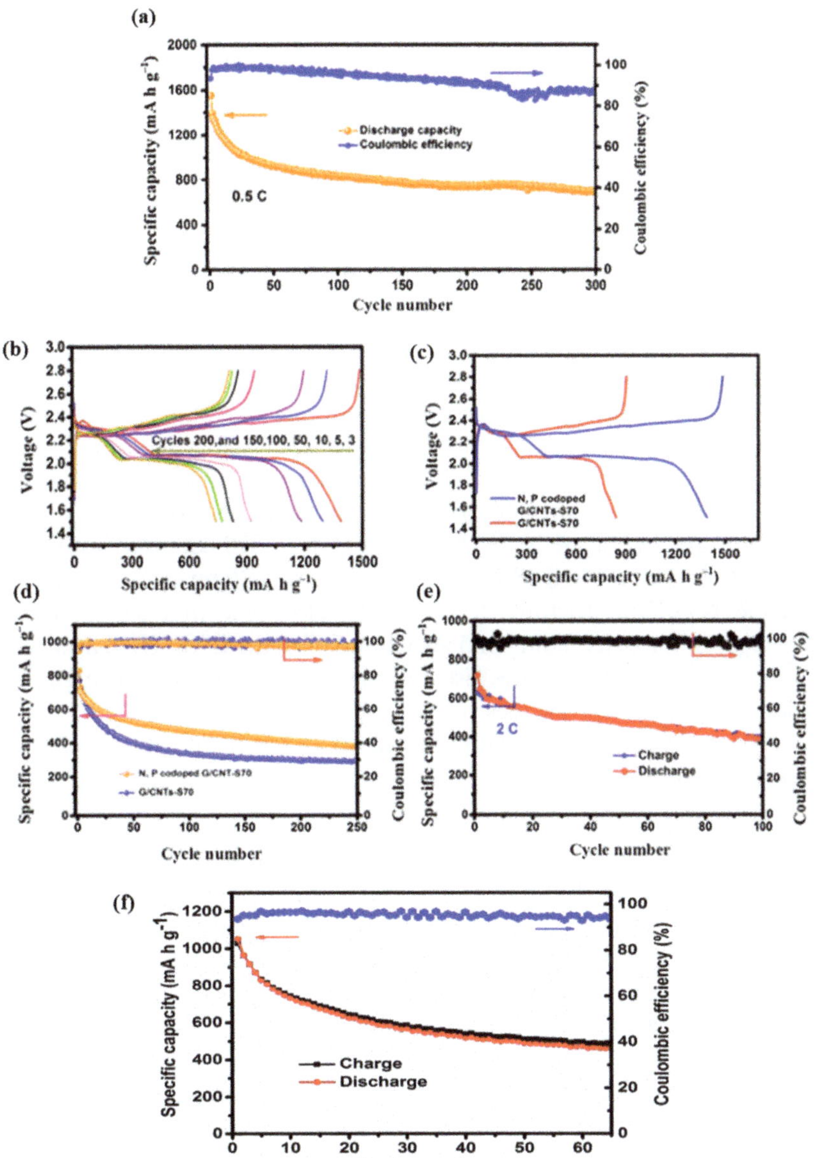

Figure 4.5. Electrochemical performance metrics of N, P co-doped G/CNTs-S70 and G/CNTs-S70 electrodes. Cycling performance of N, P co-doped G/CNTs-S70 at 0.5 C (a), discharge/charge voltage profiles of: N, P co-doped G/CNTs-S70 (b), and G/CNTs-S70(c), cycling performances of N, P co-doped G/CNTs-S70 at 1 C (d), and at 2 C (e), cycling performance and coulombic efficiency of G/CNTs-S70 for 65 cycles at 0.5 C (f). Reprinted from reference [141], copyright (2017) Springer Science Business Media, LLC, with permission from Springer Nature.

reversible capacity of 690 mAh g^{-1} over 300 cycles were demonstrated by the N, P co-doped G/CNTs-S70; while G/CNTs-S70 exhibited an initial discharge capacity of 1051 mAh g^{-1} and poor capacity retention (figure 4.5(f)). The N, P co-doped G/CNTs-S70 exhibited high coulombic efficiency, over 90% up to 215 cycles, but it declined to 86.5% over 300 cycles. The GCD profile of the N, P co-doped G/CNTs-S70 electrode is shown in figure 4.5(b), at a 0.5 C current rate with consecutive cycles. This electrode exhibits two-step discharge plateaus, which confirms the two-step interactions of sulfur with the lithium anode. In figure 4.5(c), the GCD profiles of the electrodes with N, P co-doped G/CNTs-S70 and the un-doped G/CNTs-S70 are also displayed at 0.5 C. Both graphs show two separate plateaus, as is typical for steady reduction of sulfur in LiSBs. Importantly, the N, P co-doped G/CNTs-S70 exhibits less polarization than the G/CNTs-S70 during the electrochemical reversibility reactions. At 1 C, the N, P co-doped G/CNTs-S70 and G/CNTs-S70 electrodes cycling capacities and coulombic efficiencies were compared (figure 4.5(d)). The N, P co-doped G/CNTs-S70 electrode has a greater initial capacity of 829 mAh g^{-1} and reversible capacity of 484.4 mAh g^{-1} than the G/CNTs-S70 composite, with 57.2% capacity retention. However, after 100 cycles at the same current rate of 1 C, the G/CNTs-S70 cathode's capacity decreases and lowers to 332.8 mAh g^{-1}. The coulombic efficiency for N, P co-doped G/CNTs-S70 is very high and steady at about 96.3% over 150 cycles, indicating a significant synergistic effect of N, P codoping to reduce the 'shuttle effect' and prevent the diffusion of PSs at high current rates of 1 C. Additionally, the cycling performance at a higher current rate 2 C of the N, P co-doped G/CNTs-S70 cathode showed an initial capacity of 826.1 mAh g^{-1} and 48.29% capacity retention (figure 4.5(e)). Thus, the N, P co-doped G/CNTs-S70 structure facilitated fast electronic transport, which in turn helped to suppress capacity fading and also immobilize the LPSs. High accessibility for Li ions in the hierarchical structures also contributed to the remarkable high coulombic efficiencies and reversible capacities [141].

In summary, the electrochemical performance of LiSBs based on sulfur cathodes with various carbons is due to:

- High conductivity;
- Continuous porous network that prevents the dissolution of Li polysulfide;
- Compensation for the volume expansion of the sulfur during discharge;
- Surface functional groups that enhance the polarity and conductivity of carbonaceous materials;
- Easy and quick ion diffusion and electron transport pathways;
- Strong anchoring of sulfur species through surface functionalization.

The electrochemical performance of 2D and 3D carbon derivatives including CNT, CNF, graphene, and their nanocomposites as sulfur cathodes is summarized in table 4.2 [71, 130–162].

4.6 Transition metal compounds for sulfur cathodes

By using sulfur with a low E/S ratio and efficiently reducing the core shuttling effect, LiSB performance can be improved to meet theoretical predictions. Rapid conversion of high and low-order LPSs on the surface of the host material can be a more effective approach than simply confining the PSs through physical confinement and chemical anchoring. It has been discovered that a class of materials with a flexible structure, high electrical conductivity, and polar nature is best suited for this use.

Because of their polar surface character, a number of metal-based compounds (M_aX_b, M: metal, X: anion) have been investigated as effective class sulfur host materials. In addition to strong chemical adsorption of LPSs to their surface, metal-based compounds additionally speed up the redox reaction kinetics of the adsorbed LPSs [163–165]. Metal oxides, metal sulfides, metal hydroxides, metal–organic frameworks, metal nitrides, metal carbides, and MXene are among the functional metal-based S host materials [163–168].

The investigations on the electrochemical performance of Li–S batteries employing metal oxide and metal sulfide based S cathodes, as one of the most investigated classes of metal-based materials, are highlighted in particular in the following sections.

4.6.1 Metal oxide based host materials for a sulfur cathode

Most of the transition metal oxides have polar surfaces, suitable for chemical anchoring of sulfur species though strong interaction with LPSs. The polar nature of metal oxides originate from the existence of an anion of oxygen in the O^{2-} oxidation state. Most of them also have higher redox potential which facilitates the oxidation of LPSs, beneficial for effective sulfur utilization [166, 168–170].

Through a variety of processes, including polar–polar interactions, Lewis acid–base interactions, and catalytic conversion through redox interactions, the strong interaction of a metal surface with PSs for chemical binding as well as redox conversion kinetics has been proven. Understanding the basic polysulfide-anchoring characteristics will offer new possibilities for improvement. The charge transfer reaction kinetics are hindered even with robust PSs' adsorption on polar surfaces due to the poor surface diffusion of the adsorbed LPSs, which is further reflected in the reaction's short cycle life and rate capability. According to studies, metal oxides with strong PS interactions on polar surfaces facilitate quick diffusion of ions and rapid electron transfer, display superior discharge capacities, and more cyclic stability. Therefore, the best possible equilibrium and sequential adsorption and diffusion processes on polar surfaces are required. As a result, fast electron and ion transport along conductive channels can greatly speed up redox kinetics and reduce the shuttle effect [171, 172].

Numerous metal oxides have been researched as sulfur hosts in this context. The first oxide material to be utilized as an additive in a S cathode was $Mg_{0.6}Ni_{0.4}O{\cdot}SiO_2$, investigated due to its potential to effectively trap PSs in the host matrix. Here is a review of some common metal oxides and their electrochemical performance in LiSBs as prospective S cathodes [169, 173].

TiO_2

According to reports, TiO_2 is one of the most common and promising materials for the S electrode. It exists in three crystallographic forms: anatase (A-TiO_2), rutile (R-TiO_2), and brookite (B-TiO_2) [174]. According to reports, the A-TiO_2 (101)

surface binding energy (with sulfur species) is higher (2.30 eV) than that of graphene (1.0 eV) and also higher than two other forms of TiO_2, indicating that it will perform better. Several 1D, 2D, and 3D nanostructures of TiO_2 such as mesoporous hollow TiO_2 spheres [175], TiO_2 nanofibers [176], nanoparticles [177], nanotubes [178], etc have been investigated as sulfur hosts in LiSBs.

For instance, Wang *et al* used experimental data and DFT calculations to study the performance of TiO_2 as a sulfur host. The conversion of lithium polysulfides was found to be enhanced by oxygen vacancies in TiO_{2-x}, which also boosted polysulfide-anchoring and decreased the shuttle effect. In turn, this led to a high discharge capacity, superior cyclic stability, and less capacity fading [179].

Wei *et al* prepared sulfur–TiO_2 yolk–shell composites by hydrolysis of titanium diisopropoxide bis(acetylacetonate) precursor in a colloidal sulfur suspension. The composite with a sulfur content of 53 wt% exhibited high initial specific capacity of 1030 mAh g^{-1} at 0.5 C, coulombic efficiency of 98.4%, and high capacity retention over 1000 cycles with very small capacity decay of 0.033% per cycle at 0.5 C. The enhanced performance was attributed to the available internal void in the structure, beneficial for both accommodation of volume expansion and decreasing the polysulfide dissolution. Further, the structure also assisted the electrochemical kinetics by a fast and easy ion diffusion process [180].

Liang *et al* reported a hydrogen-reduced TiO_{2-x} encapsulated sulfur composite cathode with an initial specific capacity of 1100 mAh g^{-1} at 0.2 C and good reversible specific capacity of 890 mAh g^{-1} over 200 cycles at 0.2 C. The superior performance was attributed to the 3D porous framework and hydrogen reduction, which enhances generated oxygen vacancies and electrical conductivity [181].

Additionally, it has been found that combining TiO_2 with other metal oxides or carbon materials improves the specific capacity and results in good cyclic performance through electrochemical charge transfer kinetics [182, 183]. For instance, Li *et al* synthesized a TiO_xN_y–TiO_2 quantum dots@carbon composite (TiONQDs@C) as a sulfur host. A specific discharge capacity of 869 mAh g^{-1} and a capacity retention rate of 96% after 200 cycles was reported, which is substantially superior to a carbon host sulfur cathode [184].

Xue *et al* prepared a sulfur cathode with hollow spheres of SiO_2 with a very thin coating of TiO_2 layer with high sulfur loading of about 80 wt%. The sample exhibited a specific discharge capacity of 1000 mAh g^{-1} at 1 C rate due to strong polar–polar interactions with the PSs [185]. Qian *et al* reported mesoporous TiO_2 nanotubes with a very high surface area of about 100 m^2 g^{-1} as sulfur hosts. The cathode showed specific reversible capacity of about 610 mAh g^{-1}. The consistent distribution of sulfur in the mesopores with good mechanical and electrical contact may have contributed to an increase in the charge transport rate and a good ion diffusion process [186]. Hwang *et al* fabricated a hollow TiO_2/carbon nanotubes nanocomposite for a sulfur host, which exhibited a specific capacity of 931 and 888 mAh g^{-1} at 5.0 C and 7.0 C. The electrode also showed good rate capability with high capacity retention, attributed to high electronic conductivity and channelized charge transport through the nanotubes [187]. Yu *et al* reported TiO_2 coated nitrogen-doped graphene sulfur cathode with high sulfur loading of about

80 wt%, which exhibited a high initial discharge capacity of 1069 mAh g^{-1} at 1 C and 86% capacity retention over 500 cycles at 1 C [188]. The electrochemical performance of TiO_2 with different phases and its nanocomposite as sulfur host is summarized in table 4.3 [13, 178, 188–201].

Other carbon materials, such as hollow carbon fibers, carbon aerogels, graphene, N-doped graphene, and 3D hierarchical ordered mesoporous carbon have also been studied for composites with TiO_2 [195, 198, 199, 201, 202].

Ti_4O_7

Considering the advantage of oxygen vacancies, high conductivity, and strong chemical absorption of LPSs, Magnéli-phase titanium oxide (Ti_xO_{2x-1}, e.g. TiO_2, Ti_6O_{11}, Ti_4O_7, etc) has been employed as a potential sulfur host in LiSBs [13, 203, 204].

Ti_4O_7 has been widely recognized as an excellent sulfur host in multiple recent research studies among all phases because it is at the top of the relative electrical conductivity order, which is $Ti_4O_7 > Ti_6O_{11} > TiO_2$ [205, 206].

First off, Pang *et al* reported Ti_4O_7/S composites with specific surface areas of 290 m^2g^{-1} and sulfur content of 60 wt%. These composites showed an initial specific capacity of 1069 mAh g^{-1} at 0.2 C, high capacity retention of 88% over 100 cycles, and very little capacity decay of 0.08% per cycle at 0.5 C [13].

Wei *et al* prepared nanosheet-assembled 3D mesoporous Magnéli Ti_4O_7 microspheres by an *in situ* carbonization process, which exhibited an initial specific capacity of 1317.6 mAh g^{-1} at 0.1 C. The enhanced performance was attributed to hierarchical 3D porous interconnected mesopores with large surface area, which contributed to fast kinetic processes and strong chemical interaction of Ti^{3+} with PSs [207].

Wang *et al* [208] reported an improved electrochemical performance of LiSBs with Magnéli-phase Ti_4O_7 nanoparticles and a hollow carbon spheres (HCS) core–shell nanocomposite sulfur host, synthesized by a carbothermal reduction process in an argon environment at 950 °C. To create a HCS@Ti_4O_7/S cathode electrode, the HCS@Ti_4O_7 nanocomposites were also mixed with sulfur powder in a (wt) ratio of 3:7. The pore volume of the HCS@Ti_4O_7 sulfur host was 0.58 cm^3 g^{-1}, and it had a uniformly spherical morphology with a significant specific surface area of 512 m^2 g^{-1}. The preparation and synthesis strategies for the HCS@Ti_4O_7 and HCS@Ti_4O_7/S nanocomposites are shown schematically in figure 4.6.

Figure 4.7 displays elemental mapping and scanning transmission electron microscopy-dark-field (STEM-DF) images of an HCS@Ti_4O_7/S nanocomposite electrode. It is evident that the components C, O, Ti, and S are distributed uniformly. The majority of the sulfur is localized between the inner HCS and Ti_4O_7, as seen by the overlap image in figure 4.7(f), demonstrating that the improved sulfur-trapping ability of HCS@Ti_4O_7. Ti_4O_7 nanoparticles, which have been uniformly distributed, may offer plenty of adsorption sites for polysulfides species.

The electrochemical performances of composite electrodes made of HCS@Ti_4O_7/S, HCS@TiO_2/S, and HCS are shown in figure 4.8. A cyclic voltammetry (CV) curve of HCS@Ti_4O_7 as a sulfur host was obtained at a voltage range of 1.6–2.8 V with a

Table 4.3. Electrochemical performance of metal oxides and their nanocomposite as a sulfur host.

Cathode/sulfur host	Initial specific discharge capacity mAh g^{-1}@C	Reversible capacity mAh g^{-1}@C	Cycles	Degradation rate per cycle [%]	Sulfur content [wt %]	Sulfur loading [mg cm^{-2}]	References
S-TiO_2(yolk–shell)	1030@0.5		1000	0.033	53		[180]
∝-TiO_2(anatase)	1201@0.5				48		[189]
β-TiO_2(rutile)	1135@0.5				48		[189]
γ-TiO_2(brookite)	1094@0.5				48		[189]
TiO_2(amorphous)	1030@0.5		1000		53	0.4–0.6	[190]
TiO_2	1063@0.5		500	0.04	70		[190]
TiO_{2-x}(mixed phases)	1098@0.2		200		45	0.8	[181]
TiO_2(microboxes)	605@1.0		500		49	1.0	[192]
Hierarchical TiO_2 spheres	1300@200 mA g^{-1}					0.8–1.3	[193]
TiO_2@hollow CS	1066	630@0.5	500	0.082	56	1.5	[194]
TiO_2@carbonized bacterial cellulose	1000	752@0.5	100	0.248	70	2.5	[195]
CNT@TiO_2	1258	541@0.5	1000	0.055	37	1.7	[196]
TiO_2@CNFs	1238	770@0.2	300	0.126	45	3.0	[197]
TiO_2-activated C fiber	915@0.2		100			3.0	[178]
C@TiO_2@C	774@2.0		500		61	2.5	[198]
Hollow CNFs@TiO_2	1040	650@0.5	200	0.187	54	1.6	[199]
TiO_2@G	871	732@1.0	400	0.040	44	1.0	[200]
TiO_2@Nanowire G		1053@0.2	200		62	3.2	[201]
TiO_2@N-doped G	1374@ 0.1		500		59	1.3–1.8	[188]
TiO_2@N-doped G	1069	918@1.0	500	0.028	59	1.3–1.8	[188]
Ti_4O_7	850@2.0		500		48	1.5–1.8	[13]
HCS@Ti_4O_7	1168@0.5		800		56		[208]

(*Continued*)

Table 4.3. (*Continued*)

Cathode/sulfur host	Initial specific discharge capacity mAh g^{-1}@C	Reversible capacity mAh g^{-1}@C	Cycles	Degradation rate per cycle [%]	Sulfur content [wt %]	Sulfur loading [mg cm^{-2}]	References
MnO_2	898@0.5		1500		56.5	1.7–2.1	[210]
MnO_2	1120@0.2		200	0.04	75		[210]
δ-MnO_2	1120@0.2		200		56	0.7–1.0	[211]
δ-MnO_2@S	1120@0.2	245	2000	0.036	75	0.7	[21]
MnO_2Nanosheet	0.86 mAh cm^{-2}	0.19 mAh cm^{-2}	2000@2.0		75		[21]
Core–shell S@MnO_2	780@2.0	480@2.0	800	0.048	85		[213]
Hollow S-MnO_2	644@0.5		1500	0.028			[209]
MnO_2@hollow CNF	1147	840@0.2	100	0.268	50	3.5–3.9	[210]
MnO–Ketjen CB	1059	901@1.0	200	0.075	57	1.5–2.0	[220]
MnO@porous C	475@1.6 A g^{-1}		150		52	2.8	[217]
Mn_3O_4@CNF	1055@0.1		70		50	11.0	[218]
MnO_2@hollow CB	496@4.0 A g^{-1}		200		51	0.7–1.0	[221]
Hollow CNF@MnO_2	800	593@ 1.0	400	0.065	70	2.1	[222]
MnO_2@N-doped G	1144@0.1		100			1.2	[223]
MnO_2@GO@CNTs	1025	293@1.0	2500	0.029	60–80	1.1–2.4	[224]
MnO_2@GO–CNTs	1150	964@0.2	100	0.162	64	2.8	[225]
SiO_2	937	603@0.2	200	0.178	48	1.3	[227]
SiO_2@mildly rGO	1425	763@0.1	50	0.929			[226]
Co_3O_4	1167@0.2		200	0.219	60		[228]
Co_3O_4@C	889@1.0		500		56	1.4	[229]
N-doped Co_3O_4	900@2.0		1000		60	2.13	[230]
MgO-kapok tree fibers	1368	1035@0.5	300	0.034	63–70	0.7–1.2	[171]
CeO_2@Ketjen CB	905	710@1.0	300	0.072	60		[231]
CeO_2@CB	1004	625@1.0	500	0.075	49	1.9	[232]
Nb_2O_5@mesoporpous C	1289	913@0.5	200	0.146	48	1.5	[233]

Mo_4O_{11}@G	1190	880@0.1	80	0.323	49	0.5	[234]
a-Fe_2O_3@G	670	370@2.0	500	0.090	48	0.6	[235]
C@Fe_3O_4(yolk–shell)	1104	855@0.1	200	0.113	64	5.5	[236]
ZrO_2@CNTs	1138	878@0.5	200	0.114	36		[237]
C aerogel@Nd_2O_3	1168	907@0.5	300	0.074	44.6	2.2–3.0	[238]
VO_2	1000@0.5		1000		64	1.2–1.5	[212]
V_2O_5	1000@0.2		150		64	1.2–1.5	[212]
V_2O_5	890	800@0.08	250	0.040	60	3.0	[239]
VO_2@rGO	1358@0.2		370		53	1.5	[240]
V_2O_3@C microspheres	1177	921@0.5	100	0.217	45	1.5–1.6	[241]
V_2O_5@CNFs	816	576@3	1000	0.029	52	2.0	[242]
NiO@rGO	1500	700@0.1 Ag^{-1}	150	0.355	51	4.0	[243]
NiO@rGO–Sn	1690	868@0.1 Ag^{-1}	150	0.324	47	4.0	[243]
RuO_2@mesoporous C	695	665@0.5	200	0.022	63	2.0	[244]
La_2O_3–Ketjen CB	966	720@1.0	200	0.127	42	1.5	[245]
La_2O_3@N-doped mesoporous C	1241	880@0.2	100	0.291	48		[246]
La_2O_3 Kapok tree fibers	1345	1013@0.5	300	0.047	63–70	0.7–1.2	[171]
Y_2O_3@Ketjen CB	1054	816@1.0	200	0.113	60	1.3	[247]
ZnO	949@0.2		100		55	1.0–1.2	[248]
ZnO@S@CNT			70			2.0–2.2	[249]
ZnO	1414@0.2		100		60		[250]
MgO	923@0.2		100		55	1.0–1.2	[248]
MoO_2	1100@0.1		250		32	0.32	[251]
La_2O_3	1043@1.0		100		48		[246]
Al_2O_3	750@0.5		100		64	1.0	[252]
Fe_3O_4			100			3.85	[253]
α-Fe_2O_3@G@S	1571@0.3	565	1000	0.049	60	1.0	[235]

(Continued)

Table 4.3. (*Continued*)

Cathode/sulfur host	Initial specific discharge capacity mAh g^{-1}@C	Reversible capacity mAh g^{-1}@C	Cycles	Degradation rate per cycle [%]	Sulfur content [wt %]	Sulfur loading [mg cm^{-2}]	References
$Ba_{0.5}Sr_{0.5}Co_{0.8}Fe_{0.2}O_{3-d}$@CNT	793	632@0.5	400	0.062	70	2.6–5.3	[254]
$ZnCo_2O_4$	1322@0.8 Ag^{-1}		200		50	1.1–1.3	[255]
$NiFe_2O_4$@CNTs	900@1.0	850@1.0	500	0.009	54.7	1.0–1.2	[203]
$Mg_{0.6}Ni_{0.4}O$	1545@0.1		100		61	1.0–3.0	[169]
$BaTiO_3$	1122	929@0.5	50	0.344	41	3.0	[256]
SnO_2@HCS	996	832@0.5	100	0.165	48	2.0	[257]

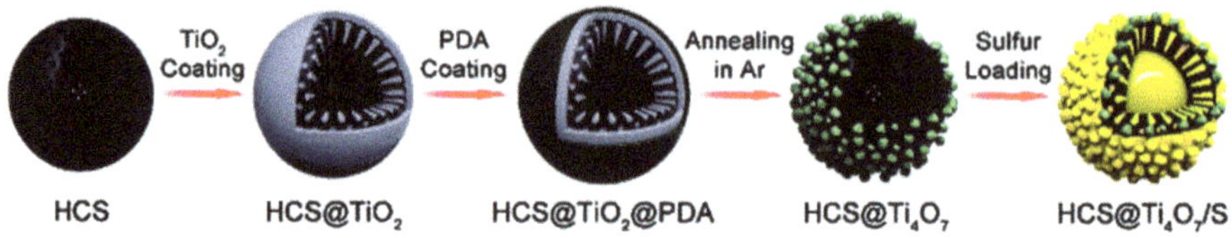

Figure 4.6. Conceptual illustration of the fabrication approach of the HCS@Ti_4O_7 sulfur host and HCS@Ti_4O_7/S nanocomposite electrode. Reproduced from [208] with permission from the Royal Society of Chemistry.

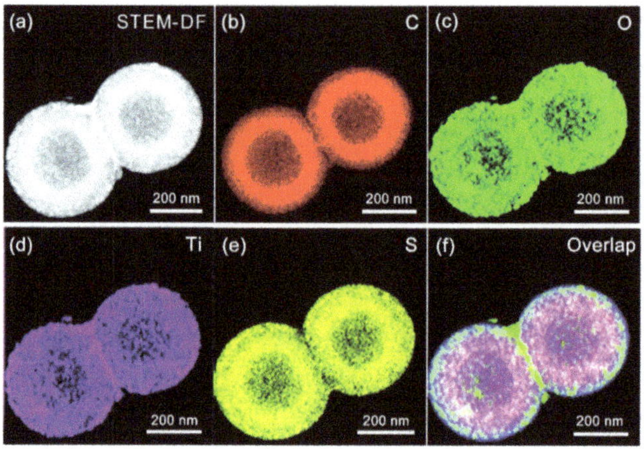

Figure 4.7. Elemental mapping and scanning transmission electron microscopy-dark-field (STEM-DF) images of a HCS@Ti_4O_7/S nanocomposite electrode. Reproduced from [208] with permission from the Royal Society of Chemistry.

scan rate of 0.05 mV s^{-1} (figure 4.8(a)). Two reduction peaks can be seen in the CV curve at 2.30 and 2.06 V, respectively. These peaks result from the reduction of S_8 to higher-order Li_2S_x and their subsequent translation to Li_2S through intermediary PSs. Additionally, the 2.38 V oxidation peak indicates a reversed transfer from Li_2S to S_8. The exceptional electrochemical reversibility of cathodic reactions is revealed by the high repeatability of the first and third cycle curves. Its highest electrical conductivity among HCS@Ti_4O_7/S, HCS@TiO_2/S, and HCS/S composite electrodes is indicated by the Nyquist plot of electrochemical impedance spectroscopy (figure 4.8(a)), which has a smaller high frequency semicircle for HCS@Ti_4O_7/S.

Figure 4.8(c) shows the GCD profiles of the HCS@Ti_4O_7/S, HCS@TiO_2/S, and HCS/S composite electrodes at 0.1 C. It reveals two discharge plateaus: one flatter plateau at 2.3 V attributed to the formation of long-chain PSs (Li_2S_x with $4 \leqslant x \leqslant 8$) from elemental sulfur, and another lower plateau at 2.1 V attributed to the formation of short-chain PSs (Li_2S_y with $2 \leqslant y \leqslant 4$, Li_2S). The cycling performance of three composite electrodes at a 0.1 C discharge rate is shown in figure 4.8(d). With a high initial discharge capacity of 1421 mAh g^{-1} and a discharge capacity of 880 mAh g^{-1} with 62% retention over 100 cycles, the HCS@Ti_4O_7/S demonstrated excellent performance. While the other two electrodes, HCS@TiO_2/S and HCS/S,

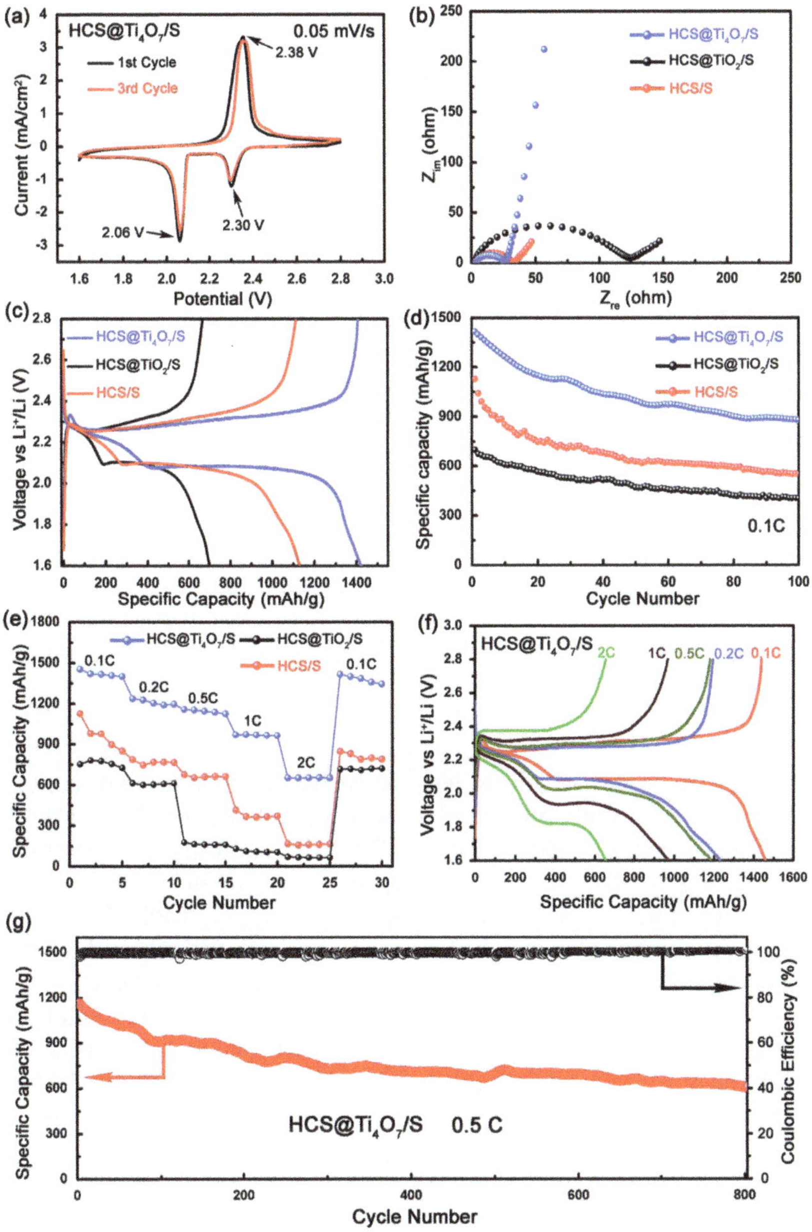

Figure 4.8. Electrochemical performance metrics of HCS@Ti_4O_7/S, HCS@TiO_2/S, and HCS/S nanocomposite electrodes: (a) CV curves; (b) Nyquist plot from the electrochemical impedance spectroscopy; (c) GCD curves; (d) cycling performance at 0.1 C; (e) rate capability; (f) GCD curves of HCS@Ti_4O_7/S electrodes at different rates; (g) cycling performance of HCS@Ti_4O_7/S electrodes for 800 cycles at 0.5 C. Reproduced from [208] with permission from the Royal Society of Chemistry.

had initial discharge capacities of 1129 and 699 mAh g^{-1}, respectively, and capacity retention of 49 and 58%.

The rate capabilities of HCS@Ti_4O_7/S were tested at various discharge rates between 0.1 C and 2 C (figure 4.8(e)) and exhibited discharge capacities of 1441, 1203, 1016, 854, and 670 mAh g^{-1} at 0.1, 0.2, 0.5, 1, and 2 C, respectively. The improvement in rate capability is the result of strong LPS chemisorption to the surface of Ti_4O_7 and a stable hollow architecture, both of which are beneficial for speedy redox kinetics and rapid ion diffusion.

The GCD patterns of HCS@Ti_4O_7/S electrodes at various current densities are also shown in figure 4.8(f). A clear discharge plateau can be detected at voltages between 1.6 and 2.8 V, even at high current densities of 2 C, illustrating the excellent stability and exceptional rate performance of the HCS@Ti_4O_7/S electrode and supporting the results in figure 4.8(e). Figure 4.8(g) illustrates the long-term performance of the HCS@Ti_4O_7/S electrode at 0.5 C. After 800 cycles, the capacity retention was still 76% and its capacity decay rate was just 0.06% per cycle. Additionally, the electrode demonstrated about 100% coulombic efficiency. The enhanced performance was attributed to physical confinement as well as strong chemical adsorption of PSs on the HCS@Ti_4O_7 surface. Moreover, the 3D hollow mesoporous network of HCS provides a highly conductive network for fast conversion of LPSs to lower order and hence suppressed the shuttle effect efficiently.

MnO_2

Manganese-based oxide, which is less expensive, abundant, and nontoxic has recently become commonly employed in secondary batteries [209–214]. The Mn ion undergoes a redox reaction with sulfur species during the electrochemical process, resulting in the creation of functional thiosulfate polythionate groups ($S_2O_3^{2-}$) on the MnO_2 surface that are helpful for trapping LPSs [212].

As of now, a number of research efforts have identified MnO_2 and its composite with carbon-based materials as a S host with better LiSB performance.

Using a variety of physical and chemical processes, including solid-state and mechano-milling, solvothermal, electro-spin, electrochemical, hydrothermal, etc, MnO_2 with the most frequent four distinct phase structures (α-, β-, γ-, δ-MnO_2) and in diverse nano architectures was effectively manufactured. The majority of them exist in non-stoichiometric form due to oxygen deficiency, which will facilitate PS chemical anchoring more effectively [21].

Liang *et al* published an XPS investigation of MnO_2 and polysulfides' interaction. Four significant peaks in the Mn 2p spectra, located at 639.4, 640.4, 641.4, and 167.2 eV were observed. The S $2p_{3/2}$ peak at 167.2 eV originates from the surface redox reaction between Li_2S_4 and δ-MnO_2 and corresponds to the thiosulfate ($[SSO_3]^{2-}$) functional group. The Mn $2p_{3/2}$ reveals a peak at 639.4, 640.4, and 641.8 eV and corresponds to oxidation states Mn^{2+} and Mn^{3+}, respectively. Thus, the confirmation of Mn reduction and existence of the thiosulfate functional group reveals the interaction between MnO_2 and PSs [21]. The MnO_2 nanosheet composite with 75 wt% sulfur showed an initial areal capacity of 0.86 mAh cm^{-2} and areal discharge capacity of 0.19 mAh cm^{-2} after 2000 cycles at 2 C. The performance was attributed to the strong chemical interaction of MnO_2 with PSs and the continuous porous structure of MnO_2, which provided conductive paths for electron transport and ion diffusion.

Liang and Nazar proposed a novel core–shell sulfur–MnO_2 composite with high sulfur loading of up to 85% as a cathode. The composite demonstrated a high initial specific capacity of 780 mAh g^{-1} at 2 C rate and maintained an increased reversible capacity of 480 mAh g^{-1} after 800 cycles at 2 C with a very low decay rate of 0.048% each cycle [213].

Wang *et al* reported MnO_2 nanosheet-decorated hollow sulfur spheres (hollow S-MnO_2) as a sulfur cathode which exhibited a discharge capacity of 644 mAh g^{-1} over 1500 cycles at 0.5 C and extremely low capacity decay rate of 0.028% per cycle. The porous design improved the chemical anchoring of the sulfur species, decreasing the effect of shuttling [209].

Manganese oxide with different structures like MnO and Mn_3O_4 has also been explored as an attractive sulfur host material due to its intrinsic polarity, beneficial for strong interaction with PSs. Additionally, its composite with conductive additives improves the kinetics of charge transfer [214–217].

For example, Chen *et al* prepared a freestanding 3D interconnected conductive network of Mn_3O_4 nanoparticles embedded in nitrogen-doped CNFs from electrospun nanofibers. The Mn_3O_4@CNF/S cathode with areal sulfur loading of 11 mg cm^{-2} showed a high initial areal capacity of 12 mAh cm^{-2} and good retention over 100 cycles. During operation, LPSs are physically confined and strongly chemically anchored by the Mn_3O_4 and N-doped CNFs, while the redox kinetics are made easier by the 3D interconnected conductive network [218].

Other metal oxides

Besides, Ti and Mn based oxides, several other oxides have been considered as sulfur host materials with the aim to improve electrochemical property and cyclic stability of LiSBs. In order to improve the overall conductivity and charge transport kinetics, the majority of metal oxides have been reported as nanocomposites with conducting additives such as nanoporous carbons, CNTs, carbon black, graphene, CNFs, and rGO. The often-studied metal oxides as sulfur hosts are MoO_3, MnO, CoO, Co_3O_4, $NiCo_2O_4$, MgO, VO_2, V_2O_3, V_2O_5, CeO_2, CaO, ZrO_2, Nb_2O_5, SnO_2, ZnO, Fe_2O_3, MoO_2, SiO_2, Al_2O_3, La_2O_3, $NiFe_2O_4$, etc [219–257].

All of these are capable of confining LPSs through strong chemical interactions and also showed enhanced electrical conductivity which accelerated the ion and charge transfer kinetics. The shuttle effect is greatly reduced by the structure engineering with carbon-based materials, resulting in both physical confinement and chemical adsorption of the sulfur species, leading to a longer cycle stability. The 3D nonporous network architecture with conductive additives such as CNTs, CNFs, rGO, etc, facilitated electron transport, resulting in substantial improvement in the electrochemical performance of LiSBs. The most important research on the electrochemical performance of LiSBs with different metal oxides and their nanocomposites is summarized in table 4.3 [169, 171, 203, 204, 210, 212, 217–257].

4.6.2 Metal sulfide based host materials for a sulfur cathode

In addition to metal oxide nanomaterials, metal sulfides have been explored as a promising new group of candidates for the S cathode. Like metal oxides, due to the polar surface, metal sulfides possess even stronger affinity for LPSs to effectively encapsulate sulfur without compromising the conductivity. Due to their large surface area and layered porous structure for easy accessibility of electroactive sites, they can exhibit improved electrochemical performance. Pyrite, spinel, and NiAs-type metal sulfides were also thoroughly investigated in addition to 2D layered transition metal disulfides as a sulfur host materials in LiSBs.

Excessive growth has been made with various abundant metal disulfides as sulfur hosts in LiSBs. The majority of current findings progressively emerged with investigations on cobalt and titanium based sulfides, due to their high electrical conductivity, ease of availability, greater number of active sites for the redox mechanism, nontoxicity, low cost, and simple processing methods. The polar Ti–S or Co–S bonds form strong chemical interactions with Li–S bonds of PSs during the charge–discharge process, inhibiting the shuttling effect. Efforts towards fundamental understanding of decomposition and oxidation of Li_2S as an effect of transition metal disulfides' affinity to LPS species were found to be a crucial step towards increased performance of LiSBs [2, 5, 7, 258].

Metal sulfides have the following advantages over other classes of materials that have aided in improving the electrochemical performance of LiSBs and increasing the usage of active sulfur [259–265]:

- Versatile material class with abundant availability;
- High electronic conductivity;
- Adjustable metallic or half-metallic characteristics and were found in three different phases including pyrite, spinel, and NiAs;
- Effective storage and quick Li-ion diffusion via 2D structures;
- Low lithiation voltages (<1.5 V vs. Li/Li^+, beyond the working potential window of LiSBs);
- Higher binding energy than metal oxides and graphene for strong chemisorption of short-intermediate to long-chain LPSs;
- Strong affinity towards polysulfide species Li_2S_x;
- Work as an activation catalyst to speed up the oxidation of Li_2S and reduce the over potential for the reverse reaction;
- Different metal oxidation states that aid in the synthesis of thiosulfate functional groups and the encapsulation of polysulfides;
- Synthesized using a range of physical and chemical synthesis methods to produce nanostructures.

The binding energy of the most often reported MS with LPSs is shown in table 4.4 [259–272]. The following part provides a quick discussion of the noteworthy developments with various TMSs as sulfur hosts in LiSBs.

Table 4.4. Binding energy of the different metal sulfides with LPSs.

Material	Binding energy (eV) from DFT calculations	Affinitive polysulfide species	References
NiS	3.6	Li_2S_4	[259]
SnS_2	3.25		[260, 261]
Bi_2S_3	2.52		[262]
NiS_2	2.06		[263]
CoS_2	1.97		[264]
Co_9S_8	1.71		[265]
SnS_2	1.26		[260, 267]
WS_2	0.8		[266]
Graphite	0.34		[264]
TiS_2	2.99	Li_2S	[268]
VS_2	2.94		[268]
Co_9S_8	2.74		[269]
ZrS_2	2.7		[268]
MoS_2	4.48		[270]
WS_2	1.45		[271]
Graphite	0.29		[272]

TiS_2

Since the invention of the S host with TMS to capture LPSs, TiS_2 has been the most commonly reported because of its strong Ti–S interactions. The theoretical prediction through *ab initio* calculations on the interaction of TiS_2 with LPSs showed strong Li–S and S–S interaction with binding energy of 2.99 eV, which is much higher than that of LPSs/graphene (0.29 eV) [181, 193, 268, 273, 274].

Since TiS_2 exhibits semi-metallic to metallic behavior in the lithiated state, it must be modified with certain conducting additives, like activated carbon, to improve the electrical contact of elemental sulfur with the surface. A decrease in capacity was seen but there was strong cycle stability when more TiS_2 was added to the active composite of the S cathode [275, 276]. Ma *et al* developed TiS_2-supported sulfur cathodes with different sulfur loadings. The electrode with 40 mg cm^{-2} showed a specific capacity of 30 mAh cm^{-2} at 2.0 mA cm^{-2} and good retention capacity of about 17 mAh cm^{-2} after 100 cycles with a capacity loss rate of 0.3% per cycle [277].

Chung *et al* fabricated a core–shell TiS_2–polysulfide composite cathode with 12 mg cm^{-2} areal sulfur loading. The hybrid cathode showed a reversible areal capacity of 10 mAh cm^{-2} after 200 cycles at 0.2 C [278]. Recently, Huang *et al* reported a TiS_2 nanosheets/S–N co-doped porous carbon (TiS_2@NSC) nanocomposite electrode using a MXene precursor. The porous carbon network that the TiS_2 nanosheets were enclosed in created a sandwich-like structure. The TiS_2@NSC electrode showed an initial areal capacity of 7.9 mAh cm^{-2} and reversible capacity of 5.9 mAh cm^{-2} over 100 cycles [279]. In all of these studies, the performance of various TiS_2-based sulfur cathodes was attributed to their superior conductivity for efficient charge transport rate, high Li-ion diffusion rate, and high binding energy between LPSs and TiS_2.

Co_9S_8

Due to their distinctive characteristics, such as polar surface nature, high electrical conductivity, being environmentally friendly and with various oxidation states, cobalt-based sulfides with various Co–S phases, such as (CoS_2, Co_3S_4, and Co_9S_8), have been employed as sulfur hosts in LiSBs more frequently than TiS_2 [265, 280–284].

The Pentlandite family Co_9S_8, which has a high electrical conductivity at room temperature of about 290 S cm^{-1} and a high binding energy of 6.06 eV between Co (008) and PSs, which is significantly higher than that of TiS_2 with a binding energy of 3.5 eV, has received the most attention among all of these [280, 281].

As an illustration, Pang *et al* prepared a graphene-like metallic Co_9S_8 nanosheet by a microwave-assisted solvothermal method as sulfur host. The S@Co_9S_8 composite electrode with a sulfur loading of about 75 wt% demonstrated a specific capacity of 1130 mAh g^{-1} and approximately 75% of the capacity retention with a very low capacity loss rate of 0.045% per cycle [265].

Later, Chen *et al* synthesized a S@Co_9S_8/hollow carbon composite electrode by a self-template method followed by sulfidation. The composite electrode showed a discharge capacity of 560 mAh g^{-1} over 1000 cycles at 2.0 C [285]. The performances are attributed to a limited shuttling effect due to the highest binding energy of 6.93 eV between LPSs and Co_9S_8, which provide strong chemisorption of PSs on its surface. Additionally, the 3D interconnected nanosheet network can contain a lot of sulfur and speed up charge transfer operations.

Moreover, another pyrite CoS_2 phase has also shown strong chemical interaction with PSs and high electrical conductivity of 6.7×10^3 S cm^{-1} [282, 283]. Yuan *et al* reported a pyrite CoS_2 and graphene material composite as a S electrode, exhibiting a high initial discharge capacity of about 1368 mAh g^{-1} with 15% CoS_2 and 75 wt% sulfur loading in the graphene host, which is about 62% higher than for a S/G cathode. The sample also demonstrated a high initial capacity of 1368 mAh g^{-1} at 0.5 C and good capacity retention with a capacity of about 1005 mAh g^{-1} over 150 cycles with very low capacity decay rate of 0.034% per cycle at 2.0 C [264].

Seo *et al* presented a composite electrode made of N-doped graphitic carbon nanoshells (NGCNs) embedded with CoS_2 nanoparticles that retained a discharge capacity of 900 mAh g^{-1} at 0.1 C over 100 cycles [286].

Zhou *et al* synthesized layer-structured N-doped carbon as a sulfur host with CoS_2 nanoparticles by carbonization of a metal–organic framework, exhibiting a specific capacity of 1326 mAh g^{-1} at 0.1 C (with 56 wt% sulfur loading) and Co/N-doped carbon exhibiting a discharge capacity of 589 mAh g^{-1} over 150 cycles [287].

He *et al* prepared a Co_9S_8 array made from a metal–organic framework (MOF) that is attached to a 3D graphene foam and serves as a sulfur host for high-performance LiSBs. The composite electrode with sulfur content of about 86.9 wt% exhibited a high areal capacity of 10.9 mAh cm^{-2} [288].

All these performances were attributed to enhanced redox kinetics owing to the notable electrical conductivity of CoS_2 and strong sulfur adsorption. According to reports, improved charge transfer mechanisms are facilitated by the strong interfacial connection between Li_2S_4 and the (111) plane of CoS_2. The composite with graphene/hollow carbon provides sufficient specific surface area and void space for sulfur infusion and volume change during the charge–discharge process. As a result of the synergistic effects of CoS_2 and porous carbon material, the composite also demonstrated improved polysulfide redox reaction kinetics in liquid phase and sought substantial sulfur consumption [264, 286, 289].

Similar to CoS_2 and Co_9S_8, Co_3O_4 has also been studied extensively as a sulfur host in LiSBs since it likewise showed strong contact with PSs, which accelerated the kinetics of the LPSs' redox process. For instance, Xu *et al* prepared a S@Co_3S_4 composite as the S cathode with areal sulfur loading of 13.5 mg cm^{-2} in hollow Co_3S_4 polyhedra. The composite exhibited a high areal capacity of 13 mAh cm^{-2} at 0.3 C [290]. Pu *et al* fabricated Co_3S_4@S nanotubes for high-performance LiSBs. These nanotubes demonstrated a specific capacity of 1267 mAh g^{-1} at 0.05 C rate as well as a very low capacity decay rate of 0.041% per cycle at 5 C over 1000 cycles. The prolonged cycle life with good capacity was attributed to the high electrical conductivity Co_3O_4 and its strong chemical affinity towards LPSs [291]. Thus, because of their inherent electrical conductivities, strong interactions with PSs, quick ion diffusion and charge transportation kinetics through hierarchically porous networks, the CoS_2, Co_3S_4, and Co_9S_8 supported composite S cathode demonstrated a better performance.

MoS_2

MoS_2, which offers strong electrical conductivity, increased binding energy to trap PSs, non-activity in the potential range of sulfur (1.8–2.6 V vs. Li/Li^+) and 2D layered structures, has been considered an appropriate choice for a sulfur host to further increase specific capacity. According to reports, the Mo-edge in the MoS_2 structure has a binding energy of around 4.48 eV compared to the sulfur-binding edge's energy of approximately 0.87 eV. As a result, it is anticipated that a sample rich in Mo will interact strongly with LPSs. The successful usage of MoS_2 and its nanocomposites with carbon materials to increase the specific surface area and hassle-free channelized path for ion diffusion has been documented in a number of research efforts [270, 292–295].

Zhao *et al* [296] designed a sulfur host, consisting of MoS_2 nanoparticles embedded in mesoporous carbon hollow spheres (MoS_2/MCHS) using a simple wet impregnation method and gas phase vulcanization method. Two composite sulfur electrodes with and without MoS_2 were prepared, termed MoS_2/MCHS/S and MCHS/S composites, respectively. A schematic of the polysulfide adsorption mechanisms for the MCHS/S and MoS_2/MCHS/S electrodes is shown in figure 4.9.

The surface morphologies for both the MoS_2/MCHS and MCHS sulfur host are shown in the highly magnified micrographs in figure 4.10(a) and (b). It can be clearly observed in the micrographs that, even after incorporation of MoS_2, MoS_2/MCHS

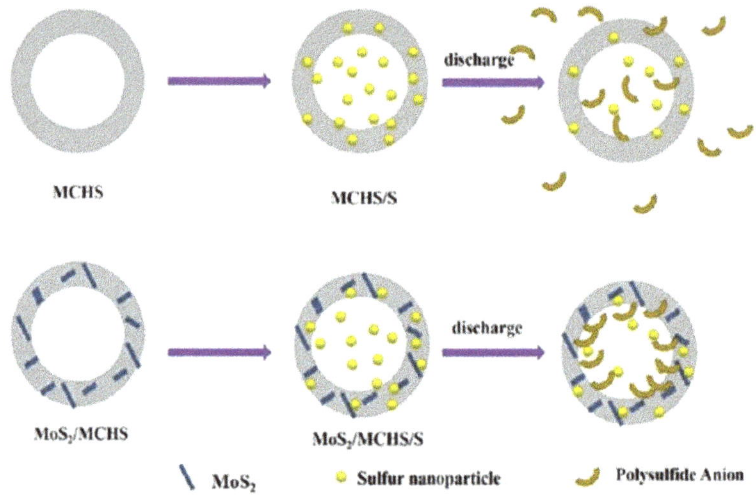

Figure 4.9. Schematic of lithium polysulfide adsorption mechanism on (a) the MCHS and (b) MoS_2/MCHS sulfur host. Adopted from [296]. Copyright 2019 Multidisciplinary Digital Publishing Institute (MDPI) under Creative Commons CC BY license.

and MCHS have a similar microporous spherical structure but a decrease in surface roughness. Figure 4.10(c) illustrates EDS element mapping, confirming the distribution of MoS_2 in the carbon shell. The C, Mo, and S elements are equally distributed throughout the hollow carbon sphere. From TEM images (figure 4.10(d)), it has been observed that the MoS_2/MCHS and MCHS structures have hollow morphologies and radial pore channels that make sulfur impregnation easy. The emendation of MoS_2 has no effect on the radial porous shells of the hollow carbon sphere, which have a thickness of about 50 nm.

Figure 4.11 shows cyclic voltammetry (CV) curves of both the MoS_2/MCHS/S and MCHS/S cathodes in the voltage range of 1.5–3.0 V at a scan rate of 0.1 mV s^{-1}. During the cathodic scan, both electrodes displayed two reduction peaks, at 2.03 and 2.32 V, resulting from the reduction of S_8 to the long-chain Li_2S_x and their further reduction to solid-state Li_2S_2/Li_2S, respectively. Subsequently, the oxidation peaks during the anodic scan at 2.49 and 2.68 V are associated with the formation of long-chain Li_2S_x, and their further continual oxidation to elemental sulfur, respectively. The high current density in the case of MoS_2/MCHS/S was attributed to increased redox kinetics of PSs as a result of their considerable chemisorption on the electrode (MoS_2/MCHS/S) surface. The potential gap between the oxidation and reduction peaks of the MoS_2/MCHS/S cathodes being lower (0.46 V) than that of the MCHS/S cathodes (0.73 V), revealed lower cell polarization and faster Li-ion diffusion in the MoS_2/MCHS/S. Using the GCD charge–discharge profiles, the specific capacity and cycle performance of both samples can be calculated.

The GCD profiles of the MoS_2/MCHS/S and MCHS/S electrodes at a range of currents from 0.1 to 4 C are shown in figure 4.12((a) and (b)). Both electrodes exhibit two discharge plateaus: one flatter plateau at 2.3 V associated with the conversion of

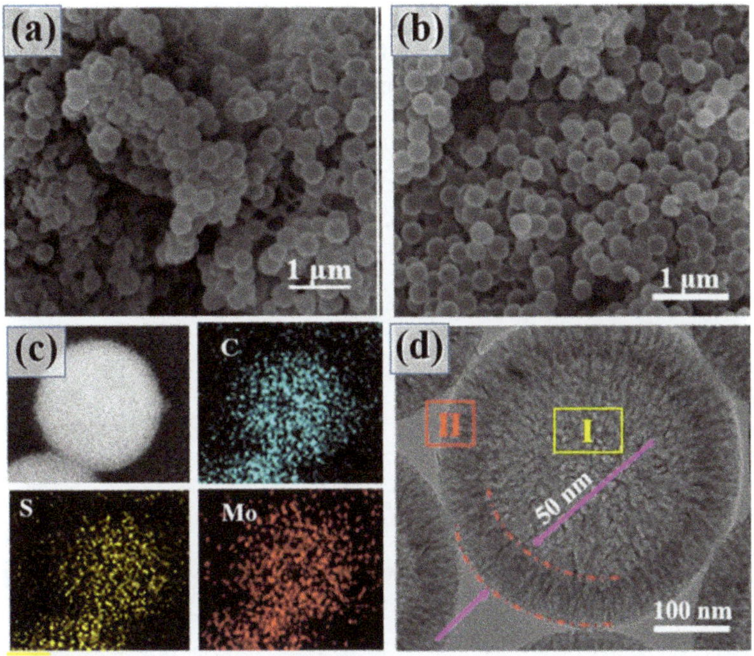

Figure 4.10. SEM micrographs of a (a) MCHS and (b) MoS2/MCHS sulfur host; (c) EDS elemental mapping images of the MoS_2/MCHS, (d) TEM image of MoS_2/MCHS. Adopted from [296]. Copyright 2019 Multidisciplinary Digital Publishing Institute (MDPI) under Creative Commons CC BY license.

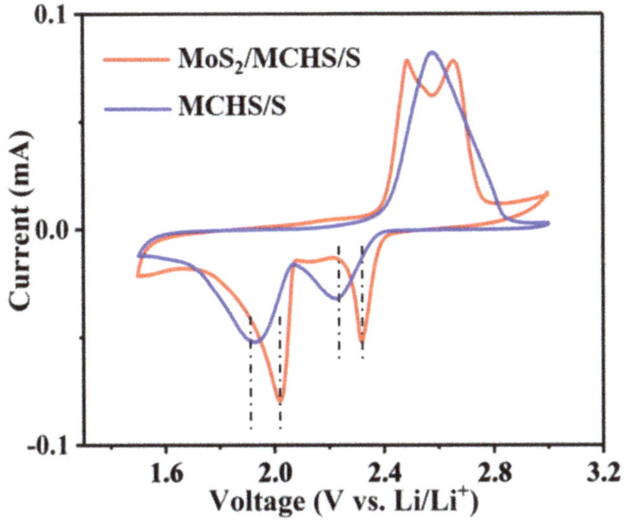

Figure 4.11. CV curves of MoS_2/MCHS/S and MCHS/S nanocomposite electrodes at a scan rate of 0.1 mV s^{-1} in the second cycle. Adopted from [296]. Copyright 2019 Multidisciplinary Digital Publishing Institute(MDPI) under Creative Commons CC BY license.

elemental sulfur to LPSs, and another lower plateau at 2.1 V associated with the formation of short-chain PSs. The shifting of peak potential towards negative (reduction peaks) or positive (oxidation peaks) in the case of MoS_2/MCHS/S compared to MCHS/S with an increase in current rate from 0.1 to 4 C has been observed and attributed to rapid polarization and electrochemical processes in the electrode. Figure 4.12((c) and (d)) shows the cycling performance of MoS_2/MCHS/S and MCHS/S at 0.2 C and 2 C. The MoS_2/MCHS/S showed a high initial discharge

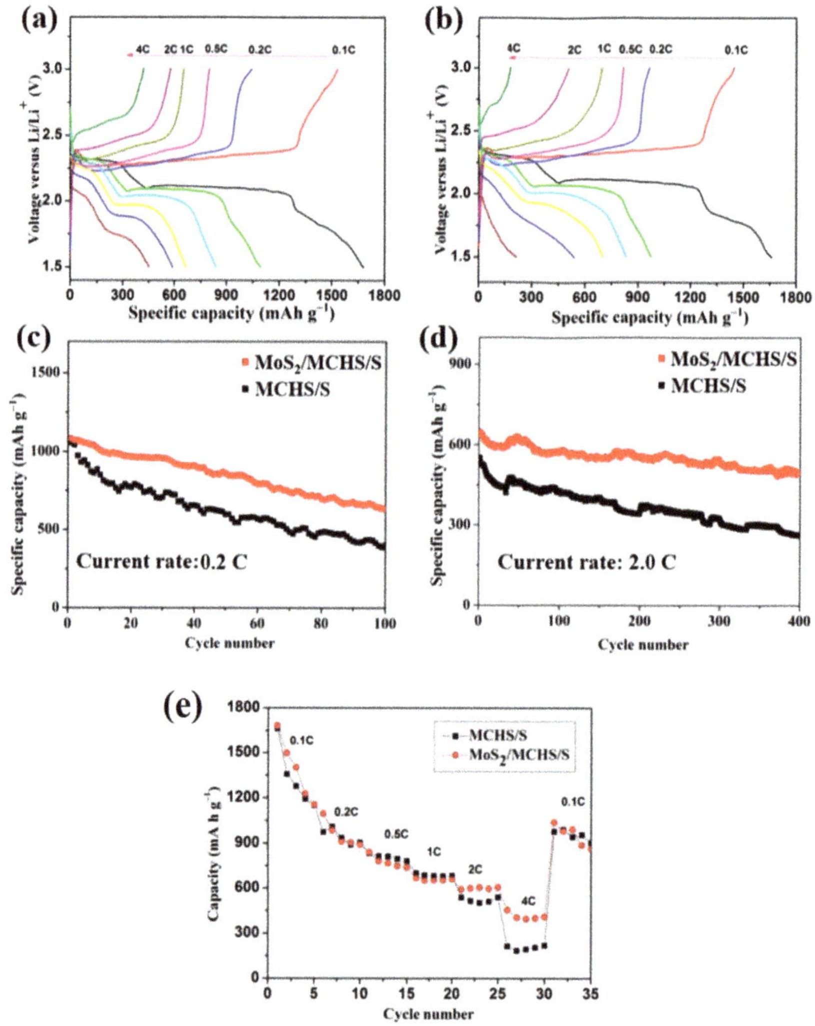

Figure 4.12. Electrochemical performance metrics: GCD curve of a MoS_2/MCHS/S electrode at various current rates from 0.1 to 4 C. (a) GCD curve of the MCHS/S electrode at various current rates from 0.1 to 4 C (b). Cycling performance of MoS_2/MCHS/S and MCHS/S nanocomposite electrodes at 0.2 C (c) and 2 C (d). Rate capabilities of MoS_2/MCHS/S and MCHS/S at a current density from 0.1 to 4 C (e). Adopted from [296]. Copyright 2019 Multidisciplinary Digital Publishing Institute (MDPI) under Creative Commons CC BY license.

capacity of 1079.5 mAh g^{-1} at 0.2 C, which exhibited gradual decrease with a decay rate of 37.40% over 100 cycles; while MCHS/S showed an initial discharge capacity of 1056.2 mAh g^{-1} at 0.2 C, which exhibited a high decay rate of 59.86% over 100 cycles. The rate capabilities of MoS_2/MCHS/S in comparison to MCHS/S were studied by increasing the charge/discharge current density from 0.1 to 4 C. MoS_2/MCHS/S has a much higher discharge capacity and rate capability than MCHS/S at various current densities (figure 4.12(e)). MoS_2/MCHS/S cathodes had discharge capacities of 1680.6, 1094.3, 838.6, 667.6, 590.2, and 453.9 mAh g^{-1} at rates of 0.1, 0.2, 0.5, 1, and 2 C, respectively. The improvement in rate capability is attributed to strong chemisorption and catalytic effect for the conversion of long-chain LPSs offered by MoS_2 and the quick charge transfer, rapid ion diffusion through a highly conductive network and a large surface area of MCHS [296]. Analyzing several other studies on MoS_2 and its nanocomposite, likewise revealed similar performance results [297–299].

Li *et al* fabricated a MWCNTs-carbon aerogel/MoS_2/sulfur composite as a sulfur cathode, which showed a discharge capacity of 1343 mAh g^{-1} at 0.2 C and capacity retention of 81% with a discharge capacity of 1088 mAh g^{-1} at 0.2 C over 200 cycles [292]. The layered structure of MoS_2 provides good electrical conductivity and also mitigated the polysulfide dissolution [300].

You *et al* synthesized 3D MoS_2/rGO foam by a hydrothermal method as a sulfur composite electrode. The porous rGO/MoS_2/sulfur composite electrode with 56% sulfur content revealed a high reversible capacity of 1100 mAh g^{-1} at 0.2 C and high reversible capacity of about 680 mAh g^{-1} at 0.5 C over 200 cycles [301].

Hu *et al* reported MoS_2 nanosheet/hierarchical carbon spheres (MoS_2@ HCS) as a sulfur host using Ni-MOF and a solvothermal process. The composite S/MoS_2@HCS showed a specific capacity of 1048 mAh g^{-1} at 0.2 C [302].

Sun *et al* reported a 3D graphene/MoS_2/sulfur composite as a sulfur cathode, which exhibited discharge capacities of 895 mAh g^{-1} at 0.2 C and 524 mAh g^{-1} at 1 C over 100 and 200 cycles, respectively [303].

The good electrochemical performances were attributed to the designed 3D porous architecture, which allows for high sulfur content, rapid Li-ion diffusion, high conductivity, high specific surface area, channelized transport of charges, and strong chemisorption of LPSs over polar MoS_2 surfaces. This accelerates the redox reaction kinetics of LPSs.

WS_2

Due to its strong electrical conductivity and active sulfiphilic edges, tungsten disulfide (WS_2) has also proven to be a reliable sulfur host. It has strong chemical interaction with lower order PSs. It has a binding energy of 1.45 eV for short-chain Li_2S species, which is advantageous for charge transfer during the redox reaction kinetics of LPSs. WS_2 and its porous carbon derivative nanocomposite have been used in a number of experiments to improve the rate capability and cycle life of LiSBs [304, 305].

For example, Babu *et al* synthesized WS_2 nanosheets as a sulfur host, revealing a higher reversible capacity of 590 mAh g^{-1} at 0.5 C with a capacity retention of 99% after 350 cycles [304].

Lei *et al* investigated the performance of 2D WS_2 nanosheets grown on cross-linked CNFs using a hydrothermal method, which demonstrated excellent discharge capacities of 1581 mAh g^{-1} at 0.1 C and 502 mAh g^{-1} at 2 C capacity retention of 90% over 1500 cycles, and a very low capacity loss rate of 0.0072% per cycle [305]. The performance was attributed to the effective adsorption of LPSs in the porous architecture and easy charge transfer pathway due to the interiors of the carbon fibres, which enhance the electrochemical activities [304, 305].

Other metal sulfides

Besides the above studies, several other metal sulfides in their different forms were investigated to significantly improve the electrochemical performance of lithium–sulfur batteries such as: CuS, FeS_2, SnS, NiS_2, VS_2, $NiCo_2S_4$, Bi_3S_3, MoS_2, ZnS, MnS, NbS_2, and so on [260–263, 270, 293, 306–313]. According to all the reports, the performance of the metal sulfide sulfur host demonstrated specific capacities between 500 and 900 mAh g^{-1} at 0.2 to 2 C. The following details were used to explain the reported results: TMS containing vacant d-orbitals, such as MnS, SnS_2, WS_2, FeS_2, MoS_2, NiS_2, and CuS, can easily interact with PSs. Some TMS, like ZrS_2 and VS_2, have shown high inherent catalytic effects that support the considerable conversion of PSs and are a result of their atomic structures. According to reports, the significant difference between the edge sites and the basal plane influences the surface shape and specific surface area in addition to improving catalytic performance.

Ye *et al* conducted a thorough examination of the chemisorption phenomenon using Raman spectroscopy. Four major NiS peaks showed a slight redshift in the analysis, indicating that NiS and PSs are chemically bonded [306]. In the same line, the study performed by Zhang and Tran on Raman spectra of FeS_2 revealed a blueshift attributed to FeS_2–Li_2S_n interactions via S–S covalent bonds [307]. For further enhancement in electrical conductivity and specific surface area, several studies on nanocomposites of TMS with carbon materials have been reported [268, 306, 308, 310, 312, 313]. The synergistic effects of the 3D porous carbon matrix and the polar surface of metal sulfides, which are beneficial for sufficient physical as well as chemical trapping of PSs and excellent electron conductive pathways, are primarily responsible for the improved electrochemical performance with a S@carbon/MS composite electrode.

Recently, Cheng *et al* [308] prepared rGO–VS_2 sheets as a sulfur host by a hydrothermal process. The rGO–VS_2/S composite electrodes were prepared by a melt-diffusion approach with different sulfur content ranging from 64, 73, and 89 wt%, corresponding to rGO–VS_2/S-64, rGO–VS_2/S-73, and rGO–VS_2/S-89, respectively.

Cyclic voltammetry (CV) curves are compared to demonstrate the overall process of sulfur oxidation and reduction. Figure 4.13(a) displays CV curves of rGO–VS_2/S-64 and rGO/S-64 in the voltage range of 1.5–3.0 V at a scan rate of 0.1 mV s^{-1}. Both electrodes showed two main reduction peaks during the cathodic scan, at 2.04 and 2.31 V, which are characteristic of long-chain Li_2S_x and their further reduction to solid-state Li_2S_2/Li_2S, respectively. The peak appeared as a weak shoulder at 1.97 V attributed to the cathodic lithiation of VS_2. Subsequently, the oxidation peaks

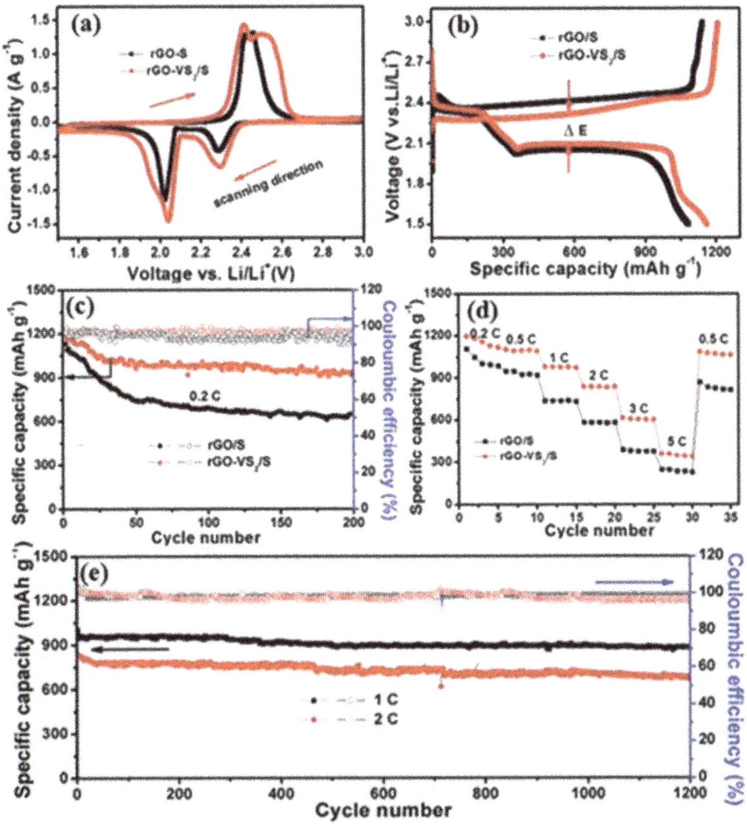

Figure 4.13. Electrochemical performance metrics of rGO/S-64 and rGO–VS_2/S-64 nanocomposite electrodes: (a) CV curves; (b) discharge/charge voltage profiles at a current rate of 0.2 C; (c) cycling performance at 0.2 C for 200 cycles; (d) the rate capability at 0.2 to 5 C; (e) cycling performance of the rGO–VS_2/S-64 electrode at 1 and 2 C over 1200 cycles. Reproduced from [308] John Wiley & Sons, copyright 2017 WILEY-VCH Verlag GmbH & Co. KGaA, Weinheim.

during the anodic scan at 2.42 and 2.55 V are associated with the formation of high-order Li_2S_x, and their further continual oxidation of elemental sulfur, respectively. The shifting of peak potential towards positive (reduction peaks)/negative (oxidation peaks) in the case of rGO–VS_2/S-64 compared to rGO/S-64, was attributed to rapid electrochemical processes in the electrode with VS_2. The high current density in the case of rGO–VS_2/S-64 was attributed to improved redox kinetics of PSs as an effect of their strong chemisorption on its surface. The superior performance of the rGO–VS_2/S-64 composite compared to the performance of rGO/S-64 was also recognized from the charge/discharge profile's constant current rate of 0.2 C (figure 4.13(b)). The GCD curve showed two discharge plateaus at 2.38 and 2.08 V, and two charge plateaus between 2.20 and 2.50 V. The rGO–VS_2/S cathode displayed longer and flatter discharge plateaus and higher capacity than rGO/S. The cycling performance of rGO–VS_2/S-64 and rGO/S-64 at 0.2 C is shown in figure 4.13(c). The rGO–VS_2/S-64 showed a high initial discharge capacity of 1194 mAh g^{-1}. The discharge capacity exhibited a gradual decrease after 27 cycles and reached 929 mAh g^{-1} after 200 cycles with an average coulombic efficiency of 98.5%.

By increasing the charge/discharge current density per five cycles from 0.2 to 5 C, the rate capabilities of rGO–VS_2/S-64 and rGO/S-64 were investigated. At different current densities, rGO–VS_2/S has a significantly higher discharge capacity and rate capability than rGO/S, as shown in figure 4.13(d). At different current densities, rGO–VS_2/S has a significantly higher discharge capacity and rate capability than rGO/S (figure 4.4(d)). Specific capacities of 1133, 987, 875, 616, and 401 mAh g^{-1} at 0.5, 1, 2, and 5 C, respectively, is obtained. To show the cycling stability performance of rGO–VS_2/S-64, the long-term cycling performance was examined and shown in figure 4.13(e). The composite electrode showed discharge capacities of 879 mAh g^{-1} at 1 C and 662 mAh g^{-1} at 2 C after 1200 cycles and good capacity retention with very low decay rates of about 0.012% and 0.017%, at 1 and 2 C, respectively. The enhanced performance is mainly attributed to the synergistic chemical and structural effects of rGO and VS_2 in the composite. The VS_2 established strong chemical interaction with LPSs and also minimizes the charge transfer resistance at the interface, while rGO provided a rapid pathway for the electron transport [308].

Table 4.5 provides a concise summary of the improvements made to LiSBs using various TMSs.

4.7 Other emerging materials as sulfur hosts for cathodes

4.7.1 Metal carbides, nitrides, and phosphides

Recently, metal carbides, nitrides, and phosphides have been investigated as promising sulfur hosts in LiSBs due to their exceptional conductivity, unique structure, and low material cost. With the introduction of some transition metal carbides in the carbon matrix as sulfur hosts, such as TiC, Fe_3C, and Mo_2C, Li–S battery performance has been significantly improved [167, 314, 315].

Metal nitrides possesses polarity, like metal carbides, and exhibit high conductivity and strong structural stability; some of these materials, including TiN and VN, have been explored as sulfur hosts with their effective affinity towards sulfur species [167, 315].

Due to the high decomposition activities of metal phosphides such as Ni_2P, Co_2P, and Fe_2P, the oxidation reaction of Li_2S was kinetically improved [316, 317]. They also accelerated the polysulfide conversion through their high conductivity, robust porous frameworks, and advantageous adsorption capabilities, giving high specific capacity and outstanding rate capability of LiSBs.

4.7.2 MXenes

A family of 2D transition metal carbides, nitrides, or carbon nitrides is called MXenes. It has the general formula, $M_{n+1}X_nT_x$, where M represents a transition metal (like Ti, Mo, Ta, Sc, V, Nb, Zr, Cr, and so on), X indicates carbon and/or nitrogen, and T_x stands for surface functional groups such as hydroxyl, oxygen, or fluorine. Since the first announcement of MXene in 2011, several MXene compositions have been synthesized successfully [318–320]. MXenes are interesting

Table 4.5. Electrochemical performance of LiSBs based on metal sulfides and their nanocomposite as sulfur hosts.

Cathode/sulfur host	Initial specific discharge capacity mAh g^{-1}@C	Cycles	Degradation rate per cycle [%]@C	Sulfur content [wt%]	Sulfur loading [mg cm^{-2}]	References
TiS_2 nanosheets	1025@0.1	100		56	7.7	[276]
Co_9S_8 host	1240@0.2		0.17@0.2	64	1.55	[280]
CoS_2 additive	1326@0.1		0.047@1.0	56	2.3	[282]
Co_9S_8 nanosheet (S@Co_9S_8)	1130	1500	0.045@0.5	75	1.5	[265]
Pyrite CoS_2@G	1368	150		75		[264]
MoS_2@C aerogel	1343@0.2	200		56	2.2	[292]
3D G@MoS_2@S	524@1.0	200				[303]
WS_2	1581@0.1		0.0072@2.0	55	2.0	[305]
2D WS_2 nanosheets	502@2.0	1500	0.0072			[304]
C@WS_2	563@2.0	1500		11	1.0–1.2	[305]
CuS QDs@C aerogel	1318@0.2	100			1.3–2.0	[313]
SnS_2	1237.5@0.2	200		48		[261]
SnS_2 additive	1400@0.1		0.058@0.5	62	2.4	[260]
Ultra-small SnS_2@HCS	1327@0.2	200		64		[261]
MoS_{2-x}@rGO	1200@0.5	600		60	1.5	[293]
MoS_2 additive	1339@0.2		0.08@0.5		2.0	[270]
Bi_2S_3	825@0.5					[262]
Bi_2S_3 additive	1480@0.1		0.028@0.5	46	2.2–3.3	[262]
NiS_2	1203@0.1		0.04@2.0	39	2.0–3.3	[263]
NiS@C	723@0.5	300		50	2.3	[306]
NiS_2-rGO	776@0.2	50		75	21	[259]
FeS_2	1129@0.16	200		62	2.0	[307]
rGO@VS_2	1130@1.0	1200		51.2	1.15	[308]
rGO@VS_2	1195@0.2	200	0.017	64		[308]
$NiCo_2S_4$	1348@0.1	1000	0.02	71		[309]

candidates for sulfur host materials because of their structural characteristics, which include high conductivity and a wealth of active surface sites [319–321]. In the meantime, new MXene structures are emerging, such as $Ti_3C_2T_x$ [322], Ti_2CT_x [323], and Mo_2CT_x [324].

4.7.3 Metal–organic frameworks

MOFs have also been reported as a very interesting class of material for energy storage devices. In MOFs, the metal ions are paired with organic linkers to form a flexible porous structure. This offers variety of coordination between the metal ions and organic linkers and their changeable chemical composition, which provides an accessible path for electrolyte penetration and substantial sulfur loading in LiSBs. The metal sites in MOFs can also act as a Lewis acid, and thus strongly trap the LPS species through both physical confinement and chemical adsorption [325–327].

4.8 Summary

The objectives of this chapter are to explore the performance of lithium–sulfur batteries with sulfur cathode based on carbon and metal compounds and to explain the role of the sulfur adsorption mechanism, various structural architectures, nanocomposites, and their composition optimization to enhance performance. High specific capacity, excellent rate capability, and long cycling life have been reported in many recent studies, achieved through a number of strategies, including hierarchical 3D porous electroconductive scaffolds, enhanced sulfur loading/content, fast charge, strong chemical adsorption of sulfur species, etc. From the said studies, it has been noted that a number of material design strategies and synthesis procedures are crucial for improving the redox kinetics. Many approaches have been proposed with several material classes for physical confinement and chemical anchoring of sulfur species. Initially, carbon-based materials with a porous structure have been considered as the most versatile materials as sulfur hosts because of their abundance, low cost, easy synthesis, high conductivity, low density, high specific surface area, and efficient physical confinement of lithium polysulfides. In a sulfur cathode, the carbon host plays three key roles: (i) physical confinement and chemically anchoring sulfur species; (ii) providing a mechanical and conductive scaffold; and (iii) increasing the amount of active surface area that is electrochemically accessible for LiSBs to function properly.

The impressive improvement in electrochemical performance with various nanoporous carbon hosts in terms of good rate capability and long life cycling stability is related to:

- A hierarchical porous structure for high sulfur storage;
- The ability to accommodate higher-order LPSs with ease in pores with desired pore sizes;
- Well-defined, organised mesoporous carbon having a higher conductivity;
- Faster electron transport through porous networks and improved ion diffusion;
- Offers efficient pathway for the numerous redox-active species.

However, it is difficult to achieve correlative pore volume and surface area even with consistent pore structure of mesoporous carbon. Also, the pore density cannot be monitored or controlled. Since the size of the higher-order LPSs is greater than that of elemental S_8, they might not fit in the micropores, which results in S cathode instability. Additionally, because the electrolyte is less accessible in very narrow micropores, sulfur usage declines.

One-dimensional carbon materials have also been used as a sulfur host and conductive additive in the cathode to improve electrical conductivity and reaction kinetics in LiSBs. CNTs and CNFs are the most often investigated 1D carbon structures for sulfur cathodes. They primarily provide a continuous 1D channel for effective charge transfer, robust cathode mechanical support, and tunable features in relation to their structural parameters. They can accommodate sulfur species both on the surface and in the hollow interior of the structure due to its hollow structure.

In order to provide a short pathway and achieve rapid ion transport, three-dimensional graphene with porous structure and appropriate pore sizes have been studied extensively in LiSBs as in other secondary batteries. Because of its high electrical conductivity, large theoretical specific surface area, high theoretical capacity, exceptional chemical durability, and inexpensive cost it causes a decrease in the diffusion path of electrolyte ions and speeds up the charge transfer.

Besides, heteroatom doping makes carbon material and graphene a better electrode due to an enhanced hydrophilic nature, increased electrical conductivity, rapid ion and charge transport in the interfaces, and significant encapsulation of LPS species via chemical adsorption as well as physical confinement.

By efficiently reducing the shuttling effect, LiSB performance can be improved to meet theoretical predictions. Instead of only confining the polysulfides through physical confinement and chemical anchoring, rapid conversion of high and low-order LPSs on the surface of the host material can be a successful strategy. It has been discovered that a class of materials with a flexible structure, high electrical conductivity, and polar nature is best suited for this use. Due to their polar adsorbent properties, a number of metal-based compounds have been investigated as effective class sulfur host materials. Strong chemical adsorption of LPS onto a metal-based compound's surface additionally accelerates the redox reaction kinetics of the adsorbed material.

A strong interaction with LPSs can chemically bind sulfur species on the polar surfaces of the majority of transition metal oxides. The majority of them exhibit improved electrical conductivity and also have greater redox potentials, which makes it easier for LPSs to be oxidized and is good for efficient sulfur use. The shuttle effect is considerably reduced by structure engineering with carbon-based materials, resulting in both physical confinement and chemical adsorption of the sulfur species, and thus prolonging the cyclic stability. The 3D nonporous network architecture improved the electrochemical performance of LiSBs significantly by facilitating electron transport.

Metal sulfides have also been investigated as a possible new category of candidates for S cathodes in addition to metal oxide nanoparticles. Due to their polar surfaces, metal sulfides have an even higher attraction for LPSs than metal oxides, effectively confining sulfur while maintaining conductivity. Due to their large surface areas, porous structure, and accessibility of electroactive sites, layered compounds can improve electrochemical performance. Pyrite, spinel, and NiAs-type metal sulfides were also thoroughly investigated in addition to 2D layered transition metal disulfides.

The enhanced electrochemical performance in metal oxides and sulfides is mostly attributed to:

- High electronic conductivity;
- Good electro catalytic activity;

- 3D electron transport pathways and continuous channels boosting the conversion of polysulfides and rate capabilities;
- Good storage capability;
- Accelerated solid–liquid–solid electrochemical reactions;
- Strong affinity towards polysulfide species Li_2S_x;
- Synergistic effects of the carbon matrix and polar surface;
- Uniformly dispersed nanoparticles in the host matrix;
- Higher binding energy than graphene for strong chemisorption of short-intermediate to long-chain LPSs;
- Various oxidation states of metal helpful for the thiosulfate functional group formation mechanism and polysulfide encapsulation mechanism;
- Accommodating the volume changes through porous structures;
- Lower overpotential for Li_2S oxidation;
- Availability in a variety of nanostructures, synthesized using several physical and chemical synthesis methods.

The multifunctional or self-sustainable hybrid systems will be of considerable interest for future development in LiSBs. Therefore, the integration of various material classes such as 3D mesoporous carbon, mixed metal oxides and sulfides, MXene, and graphene-based sulfur hosts with multidisciplinary design optimization will be a new research direction to enhance the performance of LiSBs.

References

[1] Peng H J and Zhang Q 2015 Designing host materials for sulfur cathodes: from physical confinement to surface chemistry *Angew. Chem. Int. Ed.* **54** 11018–20

[2] Yang Y, Zheng G and Cui Y 2013 Nanostructured sulfur cathodes *Chem. Soc. Rev.* **42** 3018–32

[3] Liu Z, Zheng X, Yuan N Y and Ding J N 2017 A novel device structure for a low-cost Li–S battery *J. Mater. Chem.* A **5** 942–6

[4] Lee J T, Zhao Y, Thieme S, Kim H, Oschatz M, Borchardt L, Magasinski A, Cho W I, Kaskel S and Yushin G 2013 Sulfur-infiltrated micro-and mesoporous silicon carbide-derived carbon cathode for high-performance lithium sulfur batteries *Adv. Mater.* **25** 4573–9

[5] Zhang Q, Wang Y, Seh Z W, Fu Z, Zhang R and Cui Y 2015 Understanding the anchoring effect of two-dimensional layered materials for lithium–sulfur batteries *Nano Lett.* **15** 3780–6

[6] Song X, Zhang M, Yao M, Hao C and Qiu J 2018 New insights into the anchoring mechanism of polysulfides inside nanoporous covalent organic frameworks for lithium–sulfur batteries *ACS Appl. Mater. Interfaces* **10** 43896–903

[7] Lu X *et al* 2019 High performance bimetal sulfides for lithium–sulfur batteries *Chem. Eng. J.* **358** 955–61

[8] Chen C, Li D, Gao L, Harks P P R, Eichel R A and Notten P H 2017 Carbon-coated core–shell Li_2S@C nanocomposites as high performance cathode materials for lithium–sulfur batteries *J. Mater. Chem.* A **5** 1428–33

[9] Liang J, Sun Z H, Li F and Cheng H M 2016 Carbon materials for Li–S batteries: functional evolution and performance improvement *Energy Storage Mater.* **2** 76–106

[10] Li Z, Yuan L, Yi Z, Sun Y, Liu Y, Jiang Y, Shen Y, Xin Y, Zhang Z and Huang Y 2014 Insight into the electrode mechanism in lithium-sulfur batteries with ordered microporous carbon confined sulfur as the cathode *Adv. Energy Mater.* **4** 1301473

[11] Ji X, Evers S, Black R and Nazar L F 2011 Stabilizing lithium–sulphur cathodes using polysulphide reservoirs *Nat. Commun.* **2** 1–7

[12] Wang D W, Zeng Q, Zhou G, Yin L, Li F, Cheng H M, Gentle I R and Lu G Q M 2013 Carbon–sulfur composites for Li–S batteries: status and prospects *J. Mater. Chem.* A **1** 9382–94

[13] Pang Q, Kundu D, Cuisinier M and Nazar L F 2014 Surface-enhanced redox chemistry of polysulphides on a metallic and polar host for lithium–sulphur batteries *Nat. Commun.* **5** 1–8

[14] Yao H *et al* 2014 Improving lithium–sulphur batteries through spatial control of sulphur species deposition on a hybrid electrode surface *Nat. Commun.* **5** 1–9

[15] Li C, Xi Z, Guo D, Chen X and Yin L 2018 Chemical immobilization effect on lithium polysulfides for lithium–sulfur batteries *Small* **14** 1701986

[16] Qiu Y *et al* 2014 High-rate, ultralong cycle-life lithium/sulfur batteries enabled by nitrogen-doped graphene *Nano Lett.* **14** 4821–27

[17] Liang X, Garsuch A and Nazar L F 2015 Sulfur cathodes based on conductive MXene nanosheets for high-performance lithium–sulfur batteries *Angew. Chem.* **127** 3979–83

[18] Liu C Y and Li E Y 2018 Adsorption mechanisms of lithium polysulfides on graphene-based interlayers in lithium sulfur batteries *ACS Appl. Energy Mater.* **1** 455–63

[19] Balach J, Linnemann J, Jaumann T and Giebeler L 2018 Metal-based nanostructured materials for advanced lithium–sulfur batteries *J. Mater. Chem.* A **6** 23127–68

[20] Zheng J *et al* 2014 Lewis acid–base interactions between polysulfides and metal organic framework in lithium sulfur batteries *Nano Lett.* **14** 2345–52

[21] Liang X, Hart C, Pang Q, Garsuch A, Weiss T and Nazar L F 2015 A highly efficient polysulfide mediator for lithium–sulfur batteries *Nat. Commun.* **6** 1–8

[22] Wang X, Qian Y, Wang L, Yang H, Li H, Zhao Y and Liu T 2019 Sulfurized polyacrylonitrile cathodes with high compatibility in both ether and carbonate electrolytes for ultrastable lithium–sulfur batteries *Adv. Funct. Mater.* **29** 1902929

[23] Liu M, Yang Z, Sun H, Lai C, Zhao X, Peng H and Liu T 2016 A hybrid carbon aerogel with both aligned and interconnected pores as interlayer for high-performance lithium–sulfur batteries *Nano Res.* **9** 3735–46

[24] Chung S H and Manthiram A 2014 A polyethylene glycol-supported microporous carbon coating as a polysulfide trap for utilizing pure sulfur cathodes in lithium–sulfur batteries *Adv. Mater.* **26** 7352–7

[25] Dong X, Hu N, Wei L, Su Y, Wei H, Yao L, Li X and Zhang Y 2016 A new strategy to prepare N-doped holey graphene for high-volumetric supercapacitors *J. Mater. Chem.* A **4** 9739–43

[26] Zhao M Q, Liu X F, Zhang Q, Tian G L, Huang J Q, Zhu W and Wei F 2012 Graphene/single-walled carbon nanotube hybrids: one-step catalytic growth and applications for high-rate Li–S batteries *ACS Nano* **6** 10759–69

[27] Patel N, Okabe K and Oya A 2002 Designing carbon materials with unique shapes using polymer blending and coating techniques *Carbon* **40** 315–20

[28] Hu Z, Srinivasan M P and Ni Y 2000 Preparation of mesoporous high-surface-area activated carbon *Adv. Mater.* **12** 62–5

[29] Li Y, Wang L, Gao B, Li X, Cai Q, Li Q, Peng X, Huo K and Chu P K 2017 Hierarchical porous carbon materials derived from self-template bamboo leaves for lithium–sulfur batteries *Electrochim. Acta* **229** 352–60

[30] Yushin G, Dash R, Jagiello J, Fischer J E and Gogotsi Y 2006 Carbide-derived carbons: effect of pore size on hydrogen uptake and heat of adsorption *Adv. Funct. Mater.* **16** 2288–93

[31] Hu L, Lu Y, Li X, Liang J, Huang T, Zhu Y and Qian Y 2017 Optimization of microporous carbon structures for lithium–sulfur battery applications in carbonate-based electrolyte *Small* **13** 1603533

[32] Elazari R, Salitra G, Garsuch A, Panchenko A and Aurbach D 2011 Sulfur-impregnated activated carbon fiber cloth as a binder-free cathode for rechargeable Li-S batteries *Adv. Mater.* **23** 5641–4

[33] Zhu Q, Zhao Q, An Y, Anasori B, Wang H and Xu B 2017 Ultra-microporous carbons encapsulate small sulfur molecules for high performance lithium–sulfur battery *Nano Energy* **33** 402–9

[34] Xin S, Gu L, Zhao N H, Yin Y X, Zhou L J, Guo Y G and Wan L J 2012 Smaller sulfur molecules promise better lithium–sulfur batteries *J. Am. Chem. Soc.* **134** 18510–3

[35] Li G, Sun J, Hou W, Jiang S, Huang Y and Geng J 2016 Three-dimensional porous carbon composites containing high sulfur nanoparticle content for high-performance lithium–sulfur batteries *Nat. Commun.* **7** 1–10

[36] Jayaprakash N, Shen J, Moganty S S, Corona A A L A and Archer L A 2011 Porous hollow carbon@ sulfur composites for high-power lithium–sulfur batteries *Angew. Chem. Int. Ed.* **50** 5904–8

[37] Jin L M, He F, Cai W L, Huang J X, Liu B H and Li Z P 2016 Preparation, characterization and application of modified macroporous carbon with CoN site for long-life lithium–sulfur battery *J. Power Sources* **328** 536–42

[38] Li Z, Jiang Y, Yuan L, Yi Z, Wu C, Liu Y, Strasser P and Huang Y 2014 A highly ordered meso@microporous carbon-supported sulfur@smaller sulfur core–shell structured cathode for Li–S batteries *ACS Nano* **8** 9295–303

[39] Niu S, Zhou G, Lv W, Shi H, Luo C, He Y, Li B, Yang Q H and Kang F 2016 Sulfur confined in nitrogen-doped microporous carbon used in a carbonate-based electrolyte for long-life, safe lithium–sulfur batteries *Carbon* **109** 1–6

[40] Gong Z, Wu Q, Wang F, Li X, Fan X, Yang H and Luo Z 2016 A hierarchical micro/mesoporous carbon fiber/sulfur composite for high-performance lithium–sulfur batteries *RSC Adv.* **6** 37443–51

[41] Hou Y, Li J, Gao X, Wen Z, Yuan C and Chen J 2016 3D dual-confined sulfur encapsulated in porous carbon nanosheets and wrapped with graphene aerogels as a cathode for advanced lithium sulfur batteries *Nanoscale* **8** 8228–35

[42] Zhang L, Huang H, Xia Y, Liang C, Zhang W, Luo J, Gan Y, Zhang J, Tao X and Fan H J 2017 High-content of sulfur uniformly embedded in mesoporous carbon: a new electrodeposition synthesis and an outstanding lithium–sulfur battery cathode *J. Mater. Chem.* A **5** 5905–11

[43] Oh C, Yoon N, Choi J, Choi Y, Ahn S and Lee J K 2017 Enhanced Li–S battery performance based on solution-impregnation-assisted sulfur/mesoporous carbon cathodes and a carbon-coated separator *J. Mater. Chem.* A **5** 5750–60

[44] Ji X, Lee K T and Nazar L F 2009 A highly ordered nanostructured carbon–sulphur cathode for lithium–sulphur batteries *Nat. Mater.* **8** 500–6
[45] Li X, Cao Y, Qi W, Saraf L V, Xiao J, Nie Z, Mietek J, Zhang J G, Schwenzer B and Liu J 2011 Optimization of mesoporous carbon structures for lithium–sulfur battery applications *J. Mater. Chem.* **21** 16603–10
[46] Wang D, Fu A, Li H, Wang Y, Guo P, Liu J and Zhao X S 2015 Mesoporous carbon spheres with controlled porosity for high-performance lithium–sulfur batteries *J. Power Sources* **285** 469–77
[47] Carter R, Ejorh D, Share K, Cohn A P, Douglas A, Muralidharan N, Tovar T M and Pint C L 2016 Surface oxidized mesoporous carbons derived from porous silicon as dual polysulfide confinement and anchoring cathodes in lithium sulfur batteries *J. Power Sources* **330** 70–7
[48] Liu Y, Zhao X, Chauhan G S and Ahn J H 2016 Nanostructured nitrogen-doped mesoporous carbon derived from polyacrylonitrile for advanced lithium sulfur batteries *Appl. Surf. Sci.* **380** 151–8
[49] Mi K, Chen S, Xi B, Kai S, Jiang Y, Feng J, Qian Y and Xiong S 2017 Sole chemical confinement of polysulfides on nonporous nitrogen/oxygen dual-doped carbon at the kilogram scale for lithium–sulfur batteries *Adv. Funct. Mater.* **27** 1604265
[50] Chen F, Yang J, Bai T, Long B and Zhou X 2016 Biomass waste-derived honeycomb-like nitrogen and oxygen dual-doped porous carbon for high performance lithium–sulfur batteries *Electrochim. Acta* **192** 99–109
[51] Liu S *et al* 2018 Superhierarchical cobalt-embedded nitrogen-doped porous carbon nanosheets as two-in-one hosts for high-performance lithium–sulfur batteries *Adv. Mater.* **30** 1706895
[52] Zhou X, Liao Q, Tang J, Bai T, Chen F and Yang J 2016 A high-level N-doped porous carbon nanowire modified separator for long-life lithium–sulfur batteries *J. Electroanal. Chem.* **768** 55–61
[53] Xu G, Ding B, Nie P, Shen L, Dou H and Zhang X 2014 Hierarchically porous carbon encapsulating sulfur as a superior cathode material for high performance lithium–sulfur batteries *ACS Appl. Mater. Interfaces* **6** 194–9
[54] Wang J G, Xie K and Wei B 2015 Advanced engineering of nanostructured carbons for lithium–sulfur batteries *Nano Energy* **15** 413–44
[55] Eroglu D, Zavadil K R and Gallagher K G 2015 Critical link between materials chemistry and cell-level design for high energy density and low cost lithium–sulfur transportation battery *J. Electrochem. Soc.* **162** A982
[56] Jung D S, Hwang T H, Lee J H, Koo H Y, Shakoor R A, Kahraman R, Jo Y N, Park M S and Choi J W 2014 Hierarchical porous carbon by ultrasonic spray pyrolysis yields stable cycling in lithium–sulfur battery *Nano Lett.* **14** 4418–25
[57] Liang C, Dudney N J and Howe J Y 2009 Hierarchically structured sulfur/carbon nanocomposite material for high-energy lithium battery *Chem. Mater.* **21** 4724–30
[58] Yu F, Li Y, Jia M, Nan T, Zhang H, Zhao S and Shen Q 2017 Elaborate construction and electrochemical properties of lignin-derived macro-/micro-porous carbon–sulfur composites for rechargeable lithium–sulfur batteries: the effect of sulfur-loading time *J. Alloys Compd.* **709** 677–85
[59] Lee J T, Zhao Y, Thieme S, Kim H, Oschatz M, Borchardt L, Magasinski A, Cho W I, Kaskel S and Yushin G 2013 Sulfur-infiltrated micro-and mesoporous silicon carbide-derived carbon cathode for high-performance lithium sulfur batteries *Adv. Mater.* **25** 4573–9

[60] Tang Q, Li H, Zuo M, Zhang J, Huang Y, Bai P, Xu J and Zhou K 2017 Optimized assembly of micro-/meso-/macroporous carbon for Li–S batteries *Nano* **12** 1750021

[61] Li X, Cheng X, Gao M, Ren D, Liu Y, Guo Z, Shang C, Sun L and Pan H 2017 Amylose-derived macrohollow core and microporous shell carbon spheres as sulfur host for superior lithium–sulfur battery cathodes *ACS Appl. Mater. Interfaces* **9** 10717–29

[62] He G, Evers S, Liang X, Cuisinier M, Garsuch A and Nazar L F 2013 Tailoring porosity in carbon nanospheres for lithium–sulfur battery cathodes *ACS Nano* **7** 10920–30

[63] Brun N, Sakaushi K, Yu L, Giebeler L, Eckert J and Titirici M M 2013 Hydrothermal carbon-based nanostructured hollow spheres as electrode materials for high-power lithium–sulfur batteries *Phys. Chem. Chem. Phys.* **15** 6080–7

[64] Zhou L, Huang T and Yu A 2014 Three-dimensional flower-shaped activated porous carbon/sulfur composites as cathode materials for lithium–sulfur batteries *ACS Sustain. Chem. Eng.* **2** 2442–7

[65] Rehman S, Gu X, Khan K, Mahmood N, Yang W, Huang X, Guo S and Hou Y 2016 3D vertically aligned and interconnected porous carbon nanosheets as sulfur immobilizers for high performance lithium-sulfur batteries *Adv. Energy Mater.* **6** 1502518

[66] Li G, Sun J, Hou W, Jiang S, Huang Y and Geng J 2016 Three-dimensional porous carbon composites containing high sulfur nanoparticle content for high-performance lithium–sulfur batteries *Nat. Commun.* **7** 1–10

[67] Zhang B, Qin X, Li G R and Gao X P 2010 Enhancement of long stability of sulfur cathode by encapsulating sulfur into micropores of carbon spheres *Energy Environ. Sci.* **3** 1531–7

[68] Qin X, Wu J, Xu Z L, Chong W G, Huang J Q, Liang G, Li B, Kang F and Kim J K 2019 Electrosprayed multiscale porous carbon microspheres as sulfur hosts for long-life lithium–sulfur batteries *Carbon* **141** 16–24

[69] Sun Q, He B, Zhang X Q and Lu A H 2015 Engineering of hollow core–shell interlinked carbon spheres for highly stable lithium–sulfur batteries *ACS Nano* **9** 8504–13

[70] Zhong Y, Wang S, Sha Y, Liu M, Cai R, Li L and Shao Z 2016 Trapping sulfur in hierarchically porous, hollow indented carbon spheres: a high-performance cathode for lithium–sulfur batteries *J. Mater. Chem.* A **4** 9526–35

[71] Ai W *et al* 2018 Dual confinement of polysulfides in boron-doped porous carbon sphere/graphene hybrid for advanced Li–S batteries *Nano Res.* **11** 4562–73

[72] Wang J, Yang H, Chen Z, Zhang L, Liu J, Liang P, Yang H, Shen X and Shen Z X 2018 Double-shelled phosphorus and nitrogen codoped carbon nanospheres as efficient polysulfide mediator for high-performance lithium–sulfur batteries *Adv. Sci.* **5** 1800621

[73] Li X, Tang R, Hu K, Zhang L and Ding Z 2016 Hierarchical porous carbon aerogels with VN modification as cathode matrix for high performance lithium–sulfur batteries *Electrochim. Acta* **210** 734–42

[74] Zhang Z, Li Z, Hao F, Wang X, Li Q, Qi Y, Fan R and Yin L 2014 3D interconnected porous carbon aerogels as sulfur immobilizers for sulfur impregnation for lithium-sulfur batteries with high rate capability and cycling stability *Adv. Funct. Mater.* **24** 2500–9

[75] Chang A, Wu Q, Du X, Chen S, Shen J, Song Q, Xie J and Wu W 2016 Immobilization of sulfur in microgels for lithium–sulfur battery *Chem. Commun.* **52** 4525–8

[76] Lu S, Chen Y, Zhou J, Wang Z, Wu X, Gu J, Zhang X, Pang A, Jiao Z and Jiang L 2016 A sheet-like carbon matrix hosted sulfur as cathode for high-performance lithium–sulfur batteries *Sci. Rep.* **6** 1–10

[77] Cao Z, Zhang J, Ding Y, Li Y, Shi M, Yue H, Qiao Y, Yin Y and Yang S 2016 In situ synthesis of flexible elastic N-doped carbon foam as a carbon current collector and interlayer for high-performance lithium sulfur batteries *J. Mater. Chem.* A **4** 8636–44
[78] Xu H, Qie L and Manthiram A 2016 An integrally-designed, flexible polysulfide host for high-performance lithium–sulfur batteries with stabilized lithium-metal anode *Nano Energy* **26** 224–32
[79] Gong Y, Fu C, Zhang G, Zhou H and Kuang Y 2017 Three-dimensional porous C_3N_4 nanosheets@reduced graphene oxide network as sulfur hosts for high performance lithium–sulfur batteries *Electrochim. Acta* **256** 1–9
[80] Wang J, Meng Z, Yang W, Yan X, Guo R and Han W Q 2018 Facile synthesis of rGO/g-C3N4/CNT microspheres via an ethanol-assisted spray-drying method for high-performance lithium–sulfur batteries *ACS Appl. Mater. Interfaces* **11** 819–27
[81] Pang Q, Liang X, Kwok C Y, Kulisch J and Nazar L F 2017 A comprehensive approach toward stable lithium–sulfur batteries with high volumetric energy density *Adv. Energy Mater.* **7** 1601630
[82] Zhang X Q, He B, Li W C and Lu A H 2018 Hollow carbon nanofibers with dynamic adjustable pore sizes and closed ends as hosts for high-rate lithium–sulfur battery cathodes *Nano Res.* **11** 1238–46
[83] Ma X Z, Jin B, Xin P M and Wang H H 2014 Multiwalled carbon nanotubes-sulfur composites with enhanced electrochemical performance for lithium/sulfur batteries *Appl. Surf. Sci.* **307** 346–50
[84] Kim H M, Hwang J Y, Manthiram A and Sun Y K 2016 High-performance lithium–sulfur batteries with a self-assembled multiwall carbon nanotube interlayer and a robust electrode–electrolyte interface *ACS Appl. Mater. Interfaces* **8** 983–7
[85] Liang X, Rangom Y, Kwok C Y, Pang Q and Nazar L F 2017 Interwoven MXene nanosheet/carbon-nanotube composites as Li–S cathode hosts *Adv. Mater.* **29** 1603040
[86] Patel M D, Cha E, Kang C, Gwalani B and Choi W 2017 High performance rechargeable Li–S batteries using binder-free large sulfur-loaded three-dimensional carbon nanotubes *Carbon* **118** 120–6
[87] Yuan Z, Peng H J, Huang J Q, Liu X Y, Wang D W, Cheng X B and Zhang Q 2014 Hierarchical free-standing carbon-nanotube paper electrodes with ultrahigh sulfur-loading for lithium–sulfur batteries *Adv. Funct. Mater.* **24** 6105–12
[88] Sun L *et al* 2016 Sulfur embedded in a mesoporous carbon nanotube network as a binder-free electrode for high-performance lithium–sulfur batteries *ACS Nano* **10** 1300–8
[89] Li M, Carter R, Douglas A, Oakes L and Pint C L 2017 Sulfur vapor-infiltrated 3D carbon nanotube foam for binder-free high areal capacity lithium–sulfur battery composite cathodes *ACS Nano* **11** 4877–84
[90] Han S C, Song M S, Lee H, Kim H S, Ahn H J and Lee J Y 2003 Effect of multiwalled carbon nanotubes on electrochemical properties of lithium/sulfur rechargeable batteries *J. Electrochem. Soc.* **150** A889
[91] Xin S, Gu L, Zhao N H, Yin Y X, Zhou L J, Guo Y G and Wan L J 2012 Smaller sulfur molecules promise better lithium–sulfur batteries *J. Am. Chem. Soc.* **134** 18510–3
[92] Guo J, Xu Y and Wang C 2011 Sulfur-impregnated disordered carbon nanotubes cathode for lithium–sulfur batteries *Nano Lett.* **11** 4288–94
[93] Kang H S and Sun Y K 2016 Freestanding bilayer carbon–sulfur cathode with function of entrapping polysulfide for high performance Li–S batteries *Adv. Funct. Mater.* **26** 1225–32

[94] Deng W, Hu A, Chen X, Zhang S, Tang Q, Liu Z, Fan B and Xiao K 2016 Sulfur-impregnated 3D hierarchical porous nitrogen-doped aligned carbon nanotubes as high-performance cathode for lithium–sulfur batteries *J. Power Sources* **322** 138–46
[95] Cheng X B, Huang J Q, Zhang Q, Peng H J, Zhao M Q and Wei F 2014 Aligned carbon nanotube/sulfur composite cathodes with high sulfur content for lithium–sulfur batteries *Nano Energy* **4** 65–72
[96] Su Y S, Fu Y and Manthiram A 2012 Self-weaving sulfur–carbon composite cathodes for high rate lithium–sulfur batteries *Phys. Chem. Chem. Phys.* **14** 14495–9
[97] Yuan L, Yuan H, Qiu X, Chen L and Zhu W 2009 Improvement of cycle property of sulfur-coated multi-walled carbon nanotubes composite cathode for lithium/sulfur batteries *J. Power Sources* **189** 1141–6
[98] Chen J J, Zhang Q, Shi Y N, Qin L L, Cao Y, Zheng M S and Dong Q F 2012 A hierarchical architecture S/MWCNT nanomicrosphere with large pores for lithium sulfur batteries *Phys. Chem. Chem. Phys.* **14** 5376–82
[99] Chen J, Wu D, Walter E, Engelhard M, Bhattacharya P, Pan H, Shao Y, Gao F, Xiao J and Liu J 2015 Molecular-confinement of polysulfides within mesoscale electrodes for the practical application of lithium sulfur batteries *Nano Energy* **13** 267–74
[100] Guo M Q, Huang J Q, Kong X Y, Peng H J, Shui H, Qian F Y, Zhu L, Zhu W C and Zhang Q 2016 Hydrothermal synthesis of porous phosphorus-doped carbon nanotubes and their use in the oxygen reduction reaction and lithium–sulfur batteries *New Carbon Mater.* **31** 352–62
[101] Xu D W, Xin S, You Y, Li Y, Cong H P and Yu S H 2016 Built-in carbon nanotube network inside a biomass-derived hierarchically porous carbon to enhance the performance of the sulfur cathode in a Li-S battery *ChemNanoMat* **2** 712–8
[102] Kim J H, Fu K, Choi J, Sun S, Kim J, Hu L and Paik U 2015 Hydroxylated carbon nanotube enhanced sulfur cathodes for improved electrochemical performance of lithium–sulfur batteries *Chem. Commun.* **51** 13682–5
[103] Zheng G, Yang Y, Cha J J, Hong S S and Cui Y 2011 Hollow carbon nanofiber-encapsulated sulfur cathodes for high specific capacity rechargeable lithium batteries *Nano Lett.* **11** 4462–7
[104] Rao M, Song X and Cairns E J 2012 Nano-carbon/sulfur composite cathode materials with carbon nanofiber as electrical conductor for advanced secondary lithium/sulfur cells *J. Power Sources* **205** 474–8
[105] Ji L W, Rao M M, Aloni S, Wang L, Cairns E J and Zhang Y G 2011 Porous carbon nanofiber-sulfur composite electrodes for lithium/sulfur cells *Energy Environ. Sci.* **4** 5053–9
[106] Tong Z, Huang L, Lei W, Zhang H and Zhang S 2021 Carbon-containing electrospun nanofibers for lithium–sulfur battery: Current status and future directions *J. Energy Chem.* **54** 254–73
[107] He G, Mandlmeier B, Schuster J, Nazar L F and Bein T 2014 Bimodal mesoporous carbon nanofibers with high porosity: freestanding and embedded in membranes for lithium–sulfur batteries *Chem. Mater.* **26** 3879–86
[108] Chung S H, Han P, Singhal R, Kalra V and Manthiram A 2015 Electrochemically stable rechargeable lithium–sulfur batteries with a microporous carbon nanofiber filter for polysulfide *Adv. Energy Mater.* **5** 1500738
[109] Park D, Park S and Kim D W 2021 Electrospun-cellulose derived free-standing carbon nanofibers as lightweight, ultrathin, and stackable interlayers for lithium-sulfur batteries *Chem. Eng. J.* **405** 126596

[110] Zhou F, Song L T, Lu L L, Yao H B and Yu S H 2016 Titanium-carbide-decorated carbon nanofibers as hybrid electrodes for high performance Li-S batteries *ChemNanoMat* **2** 937–41
[111] Hu F, Peng H, Zhang T, Shao W, Liu S, Wang J, Wang C and Jian X 2021 A lightweight nitrogen/oxygen dual-doping carbon nanofiber interlayer with meso-/micropores for high-performance lithium-sulfur batteries *J. Energy Chem.* **58** 115–23
[112] Xin S, Gu L, Zhao N H, Yin Y X, Zhou L J, Guo Y G and Wan L J 2012 Smaller sulfur molecules promise better lithium–sulfur batteries *J. Am. Chem. Soc.* **134** 18510–13
[113] Ye X, Ma J, Hu Y S, Wei H and Ye F 2016 MWCNT porous microspheres with an efficient 3D conductive network for high performance lithium–sulfur batteries *J. Mater. Chem.* A **4** 775–80
[114] Huang J Q, Peng H J, Liu X Y, Nie J Q, Cheng X B, Zhang Q and Wei F 2014 Flexible all-carbon interlinked nanoarchitectures as cathode scaffolds for high-rate lithium–sulfur batteries *J. Mater. Chem.* A **2** 10869–75
[115] Chen Y, Lu S, Wu X and Liu J 2015 Flexible carbon nanotube–graphene/sulfur composite film: free-standing cathode for high-performance lithium/sulfur batteries *J. Phys. Chem.* C **119** 10288–94
[116] Yun J H, Kim J H, Kim D K and Lee H W 2018 Suppressing polysulfide dissolution via cohesive forces by interwoven carbon nanofibers for high-areal-capacity lithium–sulfur batteries *Nano Lett.* **18** 475–81
[117] Li J, Guo J, Zeng L, Huang Y and Peng R 2016 Improving lithium–sulfur battery performance by using ternary hybrid cathode material *RSC Adv.* **6** 26630–36
[118] Lee J S, Kim W, Jang J and Manthiram A 2017 Sulfur-embedded activated multichannel carbon nanofiber composites for long-life, high-rate lithium–sulfur batteries *Adv. Energy Mater.* **7** 1601943
[119] Yang X, Yu Y, Yan N, Zhang H, Li X and Zhang H 2016 1-D oriented cross-linking hierarchical porous carbon fibers as a sulfur immobilizer for high performance lithium–sulfur batteries *J. Mater. Chem.* A **4** 5965–72
[120] Li Q, Zhang Z, Zhang K, Fang J, Lai Y and Li J 2014 A simple synthesis of hollow carbon nanofiber-sulfur composite via mixed-solvent process for lithium–sulfur batteries *J. Power Sources* **256** 137–44
[121] Zhou L, Lin X, Huang T and Yu A 2014 Nitrogen-doped porous carbon nanofiber webs/sulfur composites as cathode materials for lithium–sulfur batteries *Electrochim. Acta* **116** 210–6
[122] Shastri M, Shetty M, Muniyappa M, Sree M S, Gangaraju V, Sabanhalli C, Lokesh S V, Shivaramu P D and Rangappa D 2021 Reduced graphene oxide wrapped sulfur nano-composite as cathode material for lithium sulfur battery *Ceram. Int.* **47** 14790–7
[123] Luo S, Yao M, Lei S, Yan P, Wei X, Wang X, Liu L and Niu Z 2017 Freestanding reduced graphene oxide–sulfur composite films for highly stable lithium–sulfur batteries *Nanoscale* **9** 4646–51
[124] Chen K, Cao J, Lu Q, Wang Q, Yao M, Han M, Niu Z and Chen J 2018 Sulfur nanoparticles encapsulated in reduced graphene oxide nanotubes for flexible lithium–sulfur batteries *Nano Res.* **11** 1345–57
[125] Zuo P, Zhang W, Hua J, Ma Y, Du C, Cheng X, Gao Y and Yin G 2016 A novel one-dimensional reduced graphene oxide/sulfur nanoscroll material and its application in lithium sulfur batteries *Electrochim. Acta* **222** 1861–9

[126] Li N, Zheng M, Lu H, Hu Z, Shen C, Chang X, Ji G, Cao J and Shi Y 2012 High-rate lithium–sulfur batteries promoted by reduced graphene oxide coating *Chem. Commun.* **48** 4106–8
[127] Hu G, Xu C, Sun Z, Wang S, Cheng H M, Li F and Ren W 2016 3D graphene-foam–reduced-graphene-oxide hybrid nested hierarchical networks for high-performance Li–S batteries *Adv. Mater.* **28** 1603–9
[128] Chen H, Wang C, Dai Y, Qiu S, Yang J, Lu W and Chen L 2015 Rational design of cathode structure for high rate performance lithium–sulfur batteries *Nano Lett.* **15** 5443–8
[129] Zhou G, Li L, Ma C, Wang S, Shi Y, Koratkar N, Ren W, Li F and Cheng H M 2015 A graphene foam electrode with high sulfur loading for flexible and high energy Li–S batteries *Nano Energy* **11** 356–65
[130] Wang J Z, Lu L, Choucair M, Stride J A, Xu X and Liu H K 2011 Sulfur–graphene composite for rechargeable lithium batteries *J. Power Sources* **196** 7030–4
[131] Deng W, Zhou X, Fang Q and Liu Z 2016 Graphene/sulfur composites with a foam-like porous architecture and controllable pore size for high performance lithium–sulfur batteries *ChemNanoMat* **2** 952–8
[132] Fang R, Zhao S, Pei S, Qian X, Hou P X, Cheng H M, Liu C and Li F 2016 Toward more reliable lithium–sulfur batteries: an all-graphene cathode structure *ACS Nano* **10** 8676–82
[133] Zhou G, Yin L C, Wang D W, Li L, Pei S, Gentle I R, Li F and Cheng H M 2013 Fibrous hybrid of graphene and sulfur nanocrystals for high-performance lithium–sulfur batteries *ACS Nano* **7** 5367–75
[134] Pang Q, Liang X, Kwok C Y and Nazar L F 2015 The importance of chemical interactions between sulfur host materials and lithium polysulfides for advanced lithium–sulfur batteries *J. Electrochem. Soc.* **162** A2567
[135] Song J, Yu Z, Gordin M L and Wang D 2016 Advanced sulfur cathode enabled by highly crumpled nitrogen-doped graphene sheets for high-energy-density lithium–sulfur batteries *Nano Lett.* **16** 864–70
[136] Qiu Y *et al* 2014 High-rate, ultralong cycle-life lithium/sulfur batteries enabled by nitrogen-doped graphene *Nano Lett.* **14** 4821–27
[137] Huang J Q, Zhuang T Z, Zhang Q, Peng H J, Chen C M and Wei F 2015 Permselective graphene oxide membrane for highly stable and anti-self-discharge lithium–sulfur batteries *ACS Nano* **9** 3002–11
[138] Ji L, Rao M, Zheng H, Zhang L, Li Y, Duan W, Guo J, Cairns E J and Zhang Y 2011 Graphene oxide as a sulfur immobilizer in high performance lithium/sulfur cells *J. Am. Chem. Soc.* **133** 18522–5
[139] Wang X, Zhang Z, Qu Y, Lai Y and Li J 2014 Nitrogen-doped graphene/sulfur composite as cathode material for high capacity lithium–sulfur batteries *J. Power Sources* **256** 361–8
[140] Song M K, Zhang Y and Cairns E J 2013 A long-life, high-rate lithium/sulfur cell: a multifaceted approach to enhancing cell performance *Nano Lett.* **13** 5891–9
[141] Walle M D, Zhang Z, Zhang M, You X, Li Y and Liu Y N 2018 Hierarchical 3D nitrogen and phosphorous codoped graphene/carbon nanotubes–sulfur composite with synergistic effect for high performance of lithium–sulfur batteries *J. Mater. Sci.* **53** 2685–96
[142] Xu J, Shui J, Wang J, Wang M, Liu H K, Dou S X, Jeon I Y, Seo J M, Baek J B and Dai L 2014 Sulfur–graphene nanostructured cathodes via ball-milling for high-performance lithium–sulfur batteries *ACS Nano* **8** 10920–30

[143] Hu G, Xu C, Sun Z, Wang S, Cheng H M, Li F and Ren W 2016 3D graphene-foam–reduced-graphene-oxide hybrid nested hierarchical networks for high-performance Li–S batteries *Adv. Mater.* **28** 1603–9

[144] Li L, Zhou G, Yin L, Koratkar N, Li F and Cheng H M 2016 Stabilizing sulfur cathodes using nitrogen-doped graphene as a chemical immobilizer for LiS batteries *Carbon* **108** 120–6

[145] Shan J, Liu Y, Su Y, Liu P, Zhuang X, Wu D, Zhang F and Feng X 2016 Graphene-directed two-dimensional porous carbon frameworks for high-performance lithium–sulfur battery cathodes *J. Mater. Chem.* A **4** 314–20

[146] Yoo S, Lee J, Kim J M, Seong C Y, Seong K D and Piao Y 2016 Well-dispersed sulfur wrapped in reduced graphene oxide nanoscroll as cathode material for lithium–sulfur battery *J. Electroanal. Chem.* **780** 19–25

[147] Zhang Z, Wang G, Lai Y and Li J 2016 A freestanding hollow carbon nanofiber/reduced graphene oxide interlayer for high-performance lithium–sulfur batteries *J. Alloys Compd.* **663** 501–6

[148] Papandrea B *et al* 2016 Three-dimensional graphene framework with ultra-high sulfur content for a robust lithium–sulfur battery *Nano Res.* **9** 240–8

[149] Chen L, Yang R, Yan Y, Zou Y, Li X, Deng Q and Xu Y 2020 Controlling hydroxyl content of reduced graphene oxide for superior cathode performance of lithium sulfur batteries *Electrochim. Acta* **362** 137112

[150] Hou Y, Li J, Gao X, Wen Z, Yuan C and Chen J 2016 3D dual-confined sulfur encapsulated in porous carbon nanosheets and wrapped with graphene aerogels as a cathode for advanced lithium sulfur batteries *Nanoscale* **8** 8228–35

[151] Li N W, Yin Y X and Guo Y G 2016 Three-dimensional sandwich-type graphene@microporous carbon architecture for lithium–sulfur batteries *RSC Adv.* **6** 617–22

[152] Gómez-Urbano J L, Gómez-Cámer J L, Botas C, Rojo T and Carriazo D 2019 Graphene oxide-carbon nanotubes aerogels with high sulfur loadings suitable as binder-free cathodes for high performance lithiumsulfur batteries *J. Power Sources* **412** 408–15

[153] Zhang Z, Kong L L, Liu S, Li G R and Gao X P 2017 A high-efficiency sulfur/carbon composite based on 3D graphene nanosheet@carbon nanotube matrix as cathode for lithium–sulfur battery *Adv. Energy Mater.* **7** 1602543

[154] Yang H, Zhang X, Zhu W, Wang F, Li Y, Fan Q, Xiao H and Zhang F 2019 Graphene oxide induced growth of nitrogen-doped carbon nanotubes as a 1D/2D composite for high-performance lithium–sulfur batteries *ChemElectroChem* **6** 1115–21

[155] Ding Y L, Kopold P, Hahn K, Van Aken P A, Maier J and Yu Y 2016 Facile solid-state growth of 3D well-interconnected nitrogen-rich carbon nanotube–graphene hybrid architectures for lithium–sulfur batteries *Adv. Funct. Mater.* **26** 1112–9

[156] Niu S, Lv W, Zhang C, Shi Y, Zhao J, Li B, Yang Q H and Kang F 2015 One-pot self-assembly of graphene/carbon nanotube/sulfur hybrid with three dimensionally interconnected structure for lithium–sulfur batteries *J. Power Sources* **295** 182–9

[157] Liang Z L P H J, Huang J, Chen J-Q, Guo C-M, Zhu X, Li W and Zhang P Q 2015 Interconnected carbon nanotube/graphene nanosphere scaffolds as free-standing paper electrode for high-rate and ultra-stable lithium–sulfur batteries *Nano Energy* **11** 746–55

[158] Sun L, Kong W, Jiang Y, Wu H, Jiang K, Wang J and Fan S 2015 Super-aligned carbon nanotube/graphene hybrid materials as a framework for sulfur cathodes in high performance lithium sulfur batteries *J. Mater. Chem.* A **3** 5305–12

[159] Pei F, Lin L, Ou D, Zheng Z, Mo S, Fang X and Zheng N 2017 Self-supporting sulfur cathodes enabled by two-dimensional carbon yolk-shell nanosheets for high-energy-density lithium–sulfur batteries *Nat. Commun.* **8** 1–10

[160] Liu F, Chiluwal S, Childress A S, Etteh C, Miller K, Washington M, Rao A M and Podila R 2021 Graphene foam current collector for high-areal-capacity lithium–sulfur batteries *ACS Appl. Nano Mater.* **4** 53–60

[161] Fang R, Zhao S, Chen K, Wang D W and Li F 2020 Binary graphene-based cathode structure for high-performance lithium–sulfur batteries *J. Phys.: Energy* **2** 015003

[162] Cavallo C, Agostini M, Genders J P, Abdelhamid M E and Matic A 2019 A free-standing reduced graphene oxide aerogel as supporting electrode in a fluorine-free Li_2S_8 catholyte Li–S battery *J. Power Sources* **416** 111–7

[163] Liu D, Zhang C, Zhou G, Lv W, Ling G, Zhi L and Yang Q H 2018 Catalytic effects in lithium–sulfur batteries: promoted sulfur transformation and reduced shuttle effect *Adv. Sci.* **5** 1700270

[164] Zhang Z W, Peng H J, Zhao M and Huang J Q 2018 Heterogeneous/homogeneous mediators for high-energy-density lithium–sulfur batteries: progress and prospects *Adv. Funct. Mater.* **28** 1707536

[165] Yang Y, Wang Z, Jiang T, Dong C, Mao Z, Lu C, Sun W and Sun K 2018 A heterogenized Ni-doped zeolitic imidazolate framework to guide efficient trapping and catalytic conversion of polysulfides for greatly improved lithium–sulfur batteries *J. Mater. Chem.* A **6** 13593–8

[166] Liu X, Huang J Q, Zhang Q and Mai L 2017 Nanostructured metal oxides and sulfides for lithium–sulfur batteries *Adv. Mater.* **29** 1601759

[167] Cui Z, Zu C, Zhou W, Manthiram A and Goodenough J B 2016 Mesoporous titanium nitride-enabled highly stable lithium-sulfur batteries *Adv. Mater.* **28** 6926–31

[168] Rehman S, Guo S and Hou Y 2016 Rational design of Si/SiO_2@hierarchical porous carbon spheres as efficient polysulfide reservoirs for high-performance Li–S battery *Adv. Mater.* **28** 3167–72

[169] Zhang Y, Zhao Y, Yermukhambetova A, Bakenov Z and Chen P 2013 Ternary sulfur/polyacrylonitrile/$Mg_{0.6}Ni_{0.4}O$ composite cathodes for high performance lithium/sulfur batteries *J. Mater. Chem.* A **1** 295–301

[170] Fan F Y and Chiang Y M 2017 Electrodeposition kinetics in Li–S batteries: effects of low electrolyte/sulfur ratios and deposition surface composition *J. Electrochem. Soc.* **164** A917

[171] Tao X *et al* 2016 Balancing surface adsorption and diffusion of lithium-polysulfides on nonconductive oxides for lithium–sulfur battery design *Nat. Commun.* **7** 1–9

[172] Lin H, Zhang S, Zhang T, Ye H, Yao Q, Zheng G W and Lee J Y 2018 Elucidating the catalytic activity of oxygen deficiency in the polysulfide conversion reactions of lithium–sulfur batteries *Adv. Energy Mater.* **8** 1801868

[173] Tang H, Yao S, Jing M, Wu X, Hou J, Qian X, Rao D, Shen X, Xi X and Xiao K 2015 Mg0.6Ni0.4O hollow nanofibers prepared by electrospinning as additive for improving electrochemical performance of lithium–sulfur batteries *J. Alloys Compd.* **650** 351–6

[174] Yang Z, Choi D, Kerisit S, Rosso K M, Wang D, Zhang J, Graff G and Liu J 2009 Nanostructures and lithium electrochemical reactivity of lithium titanites and titanium oxides: a review *J. Power Sources* **192** 588–98

[175] Li J, Ding B, Xu G, Hou L, Zhang X and Yuan C 2013 Enhanced cycling performance and electrochemical reversibility of a novel sulfur-impregnated mesoporous hollow TiO_2 sphere cathode for advanced Li–S batteries *Nanoscale* **5** 5743–6

[176] Ma X Z, Jin B, Wang H Y, Hou J Z, Zhong X B, Wang H H and Xin P M 2015 S–TiO_2 composite cathode materials for lithium/sulfur batteries *J. Electroanal. Chem.* **736** 127–31

[177] Li Q, Zhang Z, Zhang K, Xu L, Fang J, Lai Y and Li J 2013 Synthesis and electrochemical performance of TiO_2–sulfur composite cathode materials for lithium–sulfur batteries *J. Solid State Electrochem.* **17** 2959–65

[178] Xie K, Zhang K, Han Y, Yuan K, Song Q, Wang J G, Shen C, Liu X and Wei B 2016 A novel TiO_2-wrapped activated carbon fiber/sulfur hybrid cathode for high performance lithium sulfur batteries *Electrochim. Acta* **210** 415–21

[179] Wang H E, Yin K, Qin N, Zhao X, Xia F J, Hu Z Y, Guo G, Cao G and Zhang W 2019 Oxygen-deficient titanium dioxide as a functional host for lithium–sulfur batteries *J. Mater. Chem.* A **7** 10346–53

[180] Wei Seh Z, Li W, Cha J J, Zheng G, Yang Y, McDowell M T, Hsu P C and Cui Y 2013 Sulphur–TiO_2 yolk–shell nanoarchitecture with internal void space for long-cycle lithium–sulphur batteries *Nat. Commun.* **4** 1–6

[181] Liang Z, Zheng G, Li W, Seh Z W, Yao H, Yan K, Kong D and Cui Y 2014 Sulfur cathodes with hydrogen reduced titanium dioxide inverse opal structure *ACS Nano* **8** 5249–56

[182] Shang X, Guo P, Qin T, Liu M, Lv M, Liu D and He D 2018 Sulfur immobilizer by nanoscale TiO_2 trapper deposited on hierarchical porous carbon and graphene for cathodes of lithium–sulfur batteries *Adv. Mater. Interfaces* **5** 1701602

[183] Zha C, Yang F, Zhang J, Zhang T, Dong S and Chen H 2018 Promoting polysulfide redox reactions and improving electronic conductivity in lithium–sulfur batteries via hierarchical cathode materials of graphene-wrapped porous TiO_2 microspheres with exposed (001) facets *J. Mater. Chem.* A **6** 16574–82

[184] Li H, Ma S, Li J, Liu F, Zhou H, Huang Z, Jiao S and Kuang Y 2020 Altering the reaction mechanism to eliminate the shuttle effect in lithium–sulfur batteries *Energy Storage Mater.* **26** 203–12

[185] Xue W, Yan Q B, Xu G, Suo L, Chen Y, Wang C, Wang C A and Li J 2017 Double-oxide sulfur host for advanced lithium–sulfur batteries *Nano Energy* **38** 12–8

[186] Qian X, Yang X, Jin L, Rao D, Yao S, Shen X, Xiao K, Qin S and Xiang J 2017 High rate lithium–sulfur batteries enabled by mesoporous TiO_2 nanotubes prepared by electrospinning *Mater. Res. Bull.* **95** 402–8

[187] Hwang J Y, Kim H M, Lee S K, Lee J H, Abouimrane A, Khaleel M A, Belharouak I, Manthiram A and Sun Y K 2016 High-energy, high-rate, lithium–sulfur batteries: synergetic effect of hollow TiO_2-webbed carbon nanotubes and a dual functional carbon-paper interlayer *Adv. Energy Mater.* **6** 1501480

[188] Yu M, Ma J, Song H, Wang A, Tian F, Wang Y, Qiu H and Wang R 2016 Atomic layer deposited TiO_2 on a nitrogen-doped graphene/sulfur electrode for high performance lithium–sulfur batteries *Energy Environ. Sci.* **9** 1495–503

[189] Evers S, Yim T and Nazar L F 2012 Understanding the nature of absorption/adsorption in nanoporous polysulfide sorbents for the Li–S battery *J. Phys. Chem.* C **116** 19653–8

[190] Li W, Cha J J, Zheng G, Yang Y, McDowell M T, Hsu P C and Cui Y 2013 Sulphur–TiO_2 yolk–shell nanoarchitecture with internal void space for long-cycle lithium–sulphur batteries *Nat. Commun.* **4** 1331

[191] Wang R, Wang K, Tao H, Zhao W, Jiang M, Yan J and Jiang K 2020 Controllable electrolytic formation of Ti_2O as an efficient sulfur host in lithium–sulfur (Li–S) batteries *J. Mater. Chem.* A **8** 11224–32

[192] Ni J, Jin L, Xue M, Zheng J, Zheng J P and Zhang C 2019 TiO_2 microboxes as effective polysufide reservoirs for lithium sulfur batteries *Electrochim. Acta* **296** 39–48

[193] Yang Z Z, Wang H Y, Lu L, Wang C, Zhong X B, Wang J G and Jiang Q C 2016 Hierarchical TiO_2 spheres as highly efficient polysulfide host for lithium–sulfur batteries *Sci. Rep.* **6** 1–8

[194] Li Z, Zhang J, Guan B, Wang D, Liu L M and Lou X W D 2016 A sulfur host based on titanium monoxide@ carbon hollow spheres for advanced lithium–sulfur batteries *Nat. Commun.* **7** 1–11

[195] Li F, Wang G, Wang P, Yang J, Zhang K, Liu Y and Lai Y 2017 High-performance lithium–sulfur batteries with a carbonized bacterial cellulose/TiO_2 modified separator *J. Electroanal. Chem.* **788** 150–5

[196] Yang L, Li G, Jiang X, Zhang T, Lin H and Lee J Y 2017 Balancing the chemisorption and charge transport properties of the interlayer in lithium–sulfur batteries *J. Mater. Chem.* A **5** 12506–12

[197] Zhao T *et al* 2017 A praline-like flexible interlayer with highly mounted polysulfide anchors for lithium–sulfur batteries *Small* **13** 1700357

[198] Fang M, Chen Z, Liu Y, Quan J, Yang C, Zhu L, Xu Q and Xu Q 2018 Design and synthesis of novel sandwich-type C@TiO_2@C hollow microspheres as efficient sulfur hosts for advanced lithium–sulfur batteries *J. Mater. Chem.* A **6** 1630–8

[199] Zhang Z, Li Q, Jiang S, Zhang K, Lai Y and Li J 2015 Sulfur encapsulated in a TiO_2-anchored hollow carbon nanofiber hybrid nanostructure for lithium–sulfur batteries *Chem.–A Eur. J.* **21** 1343–9

[200] Gao L, Cao M, Fu Y Q, Zhong Z, Shen Y and Wang M 2016 Hierarchical TiO_2 spheres assisted with graphene for a high performance lithium–sulfur battery *J. Mater. Chem.* A **4** 16454–61

[201] Zhou G, Zhao Y, Zu C and Manthiram A 2015 Free-standing TiO_2 nanowire-embedded graphene hybrid membrane for advanced Li/dissolved polysulfide batteries *Nano Energy* **12** 240–9

[202] Wang M, Tan S, Kan S, Wu Y, Sang S, Liu K and Liu H 2020 In-situ assembly of TiO2 with high exposure of (001) facets on three-dimensional porous graphene aerogel for lithium-sulfur battery *J. Energy Chem.* **49** 316–22

[203] Fan Q, Liu W, Weng Z, Sun Y and Wang H 2015 Ternary hybrid material for high-performance lithium–sulfur battery *J. Am. Chem. Soc.* **137** 12946–53

[204] Liu M, Jhulki S, Sun Z, Magasinski A, Hendrix C and Yushin G 2021 Atom-economic synthesis of Magnéli phase Ti_4O_7 microspheres for improved sulfur cathodes for Li–S batteries *Nano Energy* **79** 105428

[205] Mei S, Jafta C J, Lauermann I, Ran Q, Kärgell M, Ballauff M and Lu Y 2017 Porous Ti_4O_7 particles with interconnected-pore structure as a high-efficiency polysulfide mediator for lithium–sulfur batteries *Adv. Funct. Mater.* **27** 1701176

[206] Chen M, Jiang S, Huang C, Xia J, Wang X, Xiang K, Zeng P, Zhang Y and Jamil S 2018 Synergetic effects of multifunctional composites with more efficient polysulfide immobilization and ultrahigh sulfur content in lithium–sulfur batteries *ACS Appl. Mater. Interfaces* **10** 13562–72

[207] Wei H, Rodriguez E F, Best A S, Hollenkamp A F, Chen D and Caruso R A 2017 Chemical bonding and physical trapping of sulfur in mesoporous magnéli Ti_4O_7 microspheres for high-performance Li–S battery *Adv. Energy Mater.* **7** 1601616

[208] Wang F *et al* 2019 Facile synthesis of Ti_4O_7 on hollow carbon spheres with enhanced polysulfide binding for high-performance lithium–sulfur batteries *J. Mater. Chem.* A **7** 10494–504

[209] Wang X, Li G, Li J, Zhang Y, Wook A, Yu A and Chen Z 2016 Structural and chemical synergistic encapsulation of polysulfides enables ultralong-life lithium–sulfur batteries *Energy Environ. Sci.* **9** 2533–8

[210] Li Z, Zhang J and Lou X W 2015 Hollow carbon nanofibers filled with MnO_2 nanosheets as efficient sulfur hosts for lithium–sulfur batteries *Angew. Chem.* **127** 13078–82

[211] Liu S, Wang Y, Zhang N, Xun Q and Zhaofeng Z 2019 Yolk–shell MnO_2 coated sulfur particles with excellent cycle stability for Li–S batteries *Mater. Res. Express* **6** 115518

[212] Liang X *et al* 2016 Tuning transition metal oxide–sulfur interactions for long life lithium sulfur batteries: the 'Goldilocks' principle *Adv. Energy Mater.* **6** 1501636

[213] Liang X and Nazar L F 2016 In situ reactive assembly of scalable core–shell sulfur–MnO_2 composite cathodes *ACS Nano* **10** 4192–8

[214] An T, Deng D, Lei M, Wu Q H, Tian Z, Zheng M and Dong Q 2016 MnO modified carbon nanotubes as a sulfur host with enhanced performance in Li/S batteries *J. Mater. Chem.* A **4** 12858–64

[215] Zhu J, Pitcheri R, Kang T, Guo Y, Li J and Qiu Y 2018 Electrospun carbon nanofibers decorated with MnO nanoparticles as a sulfur-absorbent for lithium–sulfur batteries *Ceram. Int.* **44** 16837–43

[216] Wen X, Xiang K, Zhu Y, Xiao L, Chen X and Chen H 2018 Preparation of Mn_3O_4-CNTs microspheres as an improved sulfur hosts for lithium–sulfur batteries *Mater. Lett.* **229** 272–6

[217] Liu Y, Feng G, Guo X, Wu Z, Chen Y, Xiang W, Li J and Zhong B 2018 Employing MnO as multifunctional polysulfide reservoirs for enhanced-performance Li–S batteries *J. Alloys Compd.* **748** 100–10

[218] Chen X, Yuan L, Hao Z, Liu X, Xiang J, Zhang Z, Huang Y and Xie J 2018 Free-standing Mn_3O_4@CNF/S paper cathodes with high sulfur loading for lithium–sulfur batteries *ACS Appl. Mater. Interfaces* **10** 13406–12

[219] Luo L, Qin X, Wu J, Liang G, Li Q, Liu M, Kang F, Chen G and Li B 2018 An interwoven MoO_3@CNT scaffold interlayer for high-performance lithium–sulfur batteries *J. Mater. Chem.* A **6** 8612–9

[220] Qian X, Jin L, Zhao D, Yang X, Wang S, Shen X, Rao D, Yao S, Zhou Y and Xi X 2016 Ketjen black-MnO composite coated separator for high performance rechargeable lithium–sulfur battery *Electrochim. Acta* **192** 346–56

[221] Rehman S, Tang T, Ali Z, Huang X and Hou Y 2017 Integrated design of MnO_2@carbon hollow nanoboxes to synergistically encapsulate polysulfides for empowering lithium sulfur batteries *Small* **13** 1700087

[222] Lai Y, Wang P, Qin F, Xu M, Li J, Zhang K and Zhang Z 2017 A carbon nanofiber@mesoporous δ-MnO_2 nanosheet-coated separator for high-performance lithium–sulfur batteries *Energy Storage Mater.* **9** 179–87

[223] Yu X, Zhu X, Pei Z, Li Y, Li C and Sui Z 2021 Nitrogen-doped porous graphene/MnO_2 composite as sulfur hosts for lithium–sulfur batteries *Diam. Relat. Mater.* **118** 108497

[224] Kong W, Yan L, Luo Y, Wang D, Jiang K, Li Q, Fan S and Wang J 2017 Ultrathin MnO_2/graphene oxide/carbon nanotube interlayer as efficient polysulfide-trapping shield for high-performance Li–S batteries *Adv. Funct. Mater.* **27** 1606663

[225] Li Y, Ye D, Liu W, Shi B, Guo R, Zhao H, Pei H, Xu J and Xie J 2016 A MnO_2/graphene oxide/multi-walled carbon nanotubes-sulfur composite with dual-efficient polysulfide adsorption for improving lithium–sulfur batteries *ACS Appl. Mater. Interfaces* **8** 28566–73

[226] Campbell B, Bell J, Bay H H, Favors Z, Ionescu R, Ozkan C S and Ozkan M 2015 SiO_2-coated sulfur particles with mildly reduced graphene oxide as a cathode material for lithium–sulfur batteries *Nanoscale* **7** 7051–5

[227] Li J, Huang Y, Zhang S, Jia W, Wang X, Guo Y, Jia D and Wang L 2017 Decoration of silica nanoparticles on polypropylene separator for lithium–sulfur batteries *ACS Appl. Mater. Interfaces* **9** 7499–504

[228] Wang H, Zhou T, Li D, Gao H, Gao G, Du A, Liu H and Guo Z 2017 Ultrathin cobaltosic oxide nanosheets as an effective sulfur encapsulation matrix with strong affinity toward polysulfides *ACS Appl. Mater. Interfaces* **9** 4320–5

[229] Zhou L, Li H, Wu X, Zhang Y, Danilov D L, Eichel R A and Notten P H 2019 Double-shelled Co_3O_4/C nanocages enabling polysulfides adsorption for high-performance lithium–sulfur batteries *ACS Appl. Energy Mater.* **2** 8153–62

[230] Xu J, Zhang W, Chen Y, Fan H, Su D and Wang G 2018 MOF-derived porous N–Co_3O_4@N–C nanododecahedra wrapped with reduced graphene oxide as a high capacity cathode for lithium–sulfur batteries *J. Mater. Chem.* A **6** 2797–807

[231] Qian X, Jin L, Zhu L, Yao S, Rao D, Shen X, Xi X, Xiao K and Qin S 2016 CeO_2 nanodots decorated ketjen black for high performance lithium–sulfur batteries *RSC Adv.* **6** 111190–6

[232] Qian X, Zhao D, Jin L, Yao S, Rao D, Shen X, Zhou Y and Xi X 2016 A separator modified by spray-dried hollow spherical cerium oxide and its application in lithium sulfur batteries *RSC Adv.* **6** 114989–96

[233] Tao Y, Wei Y, Liu Y, Wang J, Qiao W, Ling L and Long D 2016 Kinetically-enhanced polysulfide redox reactions by Nb_2O_5 nanocrystals for high-rate lithium–sulfur battery *Energy Environ. Sci.* **9** 3230–9

[234] Zhou L, Ding N, Yang J, Yang L, Zong Y, Liu Z and Yu A 2016 Sulfur encapsulated in Mo_4O_{11}-anchored ultralight graphene for high-energy lithium sulfur batteries *ACS Sustain. Chem. Eng.* **4** 3679–87

[235] Zheng C *et al* 2017 Propelling polysulfides transformation for high-rate and long-life lithium–sulfur batteries *Nano Energy* **33** 306–12

[236] He J, Luo L, Chen Y and Manthiram A 2017 Yolk–shelled C@Fe_3O_4 nanoboxes as efficient sulfur hosts for high-performance lithium–sulfur batteries *Adv. Mater.* **29** 1702707

[237] Zhou Y, Zhou C, Li Q, Yan C, Han B, Xia K, Gao Q and Wu J 2015 Enabling prominent high-rate and cycle performances in one lithium–sulfur battery: designing permselective gateways for Li^+ transportation in Holey-CNT/S cathodes *Adv. Mater.* **27** 3774–81

[238] Li X, Zhang L, Ding Z and He Y 2017 Ultrafine Nd_2O_3 nanoparticles doped carbon aerogel to immobilize sulfur for high performance lithium–sulfur batteries *J. Electroanal. Chem.* **799** 617–24

[239] Li W, Hicks-Garner J, Wang J, Liu J, Gross A F, Sherman E, Graetz J, Vajo J J and Liu P 2014 V_2O_5 polysulfide anion barrier for long-lived Li–S batteries *Chem. Mater.* **26** 3403–10

[240] Li S, Cen Y, Xiang Q, Aslam M K, Hu B, Li W, Tang Y, Yu Q, Liu Y and Chen C 2019 Vanadium dioxide–reduced graphene oxide binary host as an efficient polysulfide plague for high-performance lithium–sulfur batteries *J. Mater. Chem.* A **7** 1658–68

[241] Tang R, Li X, Ding Z and Zhang L 2016 An ultrafine V_2O_3 modified hierarchical porous carbon microsphere as a high performance cathode matrix for lithium–sulfur batteries *RSC Adv.* **6** 65162–70

[242] Liu M, Li Q, Qin X, Liang G, Han W, Zhou D, He Y B, Li B and Kang F 2017 Suppressing self-discharge and shuttle effect of lithium–sulfur batteries with V_2O_5-decorated carbon nanofiber interlayer *Small* **13** 1602539

[243] Li C, Dong S, Guo D, Zhang Z, Wang M and Yin L 2017 Ternary NiO/RGO-Sn hybrid flexible freestanding film as interlayer for lithium–sulfur batteries with improved performance *Electrochim. Acta* **251** 43–50

[244] Balach J, Jaumann T, Mühlenhoff S, Eckert J and Giebeler L 2016 Enhanced polysulphide redox reaction using a RuO_2 nanoparticle-decorated mesoporous carbon as functional separator coating for advanced lithium–sulphur batteries *Chem. Commun.* **52** 8134–7

[245] Qian X, Zhao D, Jin L, Shen X, Yao S, Rao D, Zhou Y and Ming Xi X 2017 Hollow spherical Lanthanum oxide coated separator for high electrochemical performance lithium–sulfur batteries *Mater. Res. Bull.* **94** 104–12

[246] Sun F, Wang J, Long D, Qiao W, Ling L, Lv C and Cai R 2013 A high-rate lithium–sulfur battery assisted by nitrogen-enriched mesoporous carbons decorated with ultrafine La_2O_3 nanoparticles *J. Mater. Chem.* A **1** 13283–9

[247] Wang S, Qian X, Jin L, Rao D, Yao S, Shen X, Xiao K and Qin S 2017 Separator modified by Y_2O_3 nanoparticles-Ketjen Black hybrid and its application in lithium–sulfur battery *J. Solid State Electrochem.* **21** 3229–36

[248] Yu M, Wang A, Tian F, Song H, Wang Y, Li C, Hong J D and Shi G 2015 Dual-protection of a graphene–sulfur composite by a compact graphene skin and an atomic layer deposited oxide coating for a lithium–sulfur battery *Nanoscale* **7** 5292–8

[249] Gu X, Tong C J, Wen B, Liu L M, Lai C and Zhang S 2016 Ball-milling synthesis of ZnO@sulphur/carbon nanotubes and $Ni(OH)_2$@sulphur/carbon nanotubes composites for high-performance lithium–sulphur batteries *Electrochim. Acta* **196** 369–76

[250] Liang X, Song Q, Liu Y and Liu H 2015 Preparation of ZnO porous nanostructures and its application in cathode material for lithium sulfur battery *Int. J. Electrochem. Sci.* **10** 9333–41

[251] Qu Q, Gao T, Zheng H, Wang Y, Li X, Li X, Chen J, Han Y, Shao J and Zheng H 2015 Strong surface-bound sulfur in conductive MoO_2 matrix for enhancing Li–S battery performance *Adv. Mater. Interfaces* **2** 1500048

[252] Yu M, Yuan W, Li C, Hong J D and Shi G 2014 Performance enhancement of a graphene–sulfur composite as a lithium–sulfur battery electrode by coating with an ultrathin Al_2O_3 film via atomic layer deposition *J. Mater. Chem.* A **2** 7360–6

[253] Ding M, Huang S, Wang Y, Hu J, Pam M E, Fan S, Shi Y, Ge Q and Yang H Y 2019 Promoting polysulfide conversion by catalytic ternary Fe_3O_4/carbon/graphene composites with ordered microchannels for ultrahigh-rate lithium–sulfur batteries *J. Mater. Chem.* A **7** 25078–87

[254] Kong L, Chen X, Li B Q, Peng H J, Huang J Q, Xie J and Zhang Q 2018 A bifunctional perovskite promoter for polysulfide regulation toward stable lithium–sulfur batteries *Adv. Mater.* **30** 1705219

[255] Sun Q, Xi B, Li J Y, Mao H, Ma X, Liang J, Feng J and Xiong S 2018 Nitrogen-doped graphene-supported mixed transition-metal oxide porous particles to confine polysulfides for lithium–sulfur batteries *Adv. Energy Mater.* **8** 1800595

[256] Yim T *et al* 2016 Effective polysulfide rejection by dipole-aligned $BaTiO_3$ coated separator in lithium–sulfur batteries *Adv. Funct. Mater.* **26** 7817–23

[257] Liu J, Yuan L, Yuan K, Li Z, Hao Z, Xiang J and Huang Y 2016 SnO_2 as a high-efficiency polysulfide trap in lithium–sulfur batteries *Nanoscale* **8** 13638–45

[258] Zhou G *et al* 2017 Catalytic oxidation of Li_2S on the surface of metal sulfides for Li–S batteries *Proc. Natl. Acad. Sci.* 114 840–5

[259] Luo L, Chung S H and Manthiram A 2018 Rational design of a dual-function hybrid cathode substrate for lithium–sulfur batteries *Adv. Energy Mater.* **8** 1801014

[260] Li X, Lu Y, Hou Z, Zhang W, Zhu Y, Qian Y, Liang J and Qian Y 2016 SnS2-compared to SnO2-stabilized S/C composites toward high-performance lithium sulfur batteries *ACS Appl. Mater. Interfaces* **8** 19550–7

[261] Li X, Chu L, Wang Y and Pan L 2016 Anchoring function for polysulfide ions of ultrasmall SnS_2 in hollow carbon nanospheres for high performance lithium–sulfur batteries *Mater. Sci. Eng.* **B 205** 46–54

[262] Li X, Liang J, Lu Y, Hou Z, Zhang W, Zhu Y and Qian Y 2016 Bi_2S_3 *in situ* formed in molten S environment stabilized sulfur cathodes for high-performance lithium–sulfur batteries *J. Power Sources* **329** 379–86

[263] Lu Y, Li X, Liang J, Hu L, Zhu Y and Qian Y 2016 A simple melting-diffusing-reacting strategy to fabricate S/NiS_2–C for lithium–sulfur batteries *Nanoscale* **8** 17616–22

[264] Yuan Z, Peng H J, Hou T Z, Huang J Q, Chen C M, Wang D W, Cheng X B, Wei F and Zhang Q 2016 Powering lithium–sulfur battery performance by propelling polysulfide redox at sulfiphilic hosts *Nano Lett.* **16** 519–27

[265] Pang Q, Kundu D and Nazar L F 2016 A graphene-like metallic cathode host for long-life and high-loading lithium–sulfur batteries *Mater. Horiz.* **3** 130–6

[266] Ki W, Huang X, Li J, Young D L and Zhang Y 2007 Highly conductive group VI transition metal dichalcogenide films by solution-processed deposition *J. Mater. Res.* **22** 1390–5

[267] Voznyi A, Kosyak V, Opanasyuk A, Tirkusova N, Grase L, Medvids A and Mezinskis G 2016 Structural and electrical properties of SnS_2 thin films *Mater. Chem. Phys.* **173** 52–61

[268] Seh Z W, Yu J H, Li W, Hsu P C, Wang H, Sun Y, Yao H, Zhang Q and Cui Y 2014 Two-dimensional layered transition metal disulphides for effective encapsulation of high-capacity lithium sulphide cathodes *Nat. Commun.* **5** 1–8

[269] Kumar N, Raman N and Sundaresan A 2014 Synthesis and properties of cobalt sulfide phases: CoS_2 and Co_9S_8 *Z. Anorg. Allg. Chem.* **640** 1069–74

[270] Wang H, Zhang Q, Yao H, Liang Z, Lee H W, Hsu P C, Zheng G and Cui Y 2014 High electrochemical selectivity of edge versus terrace sites in two-dimensional layered MoS_2 materials *Nano Lett.* **14** 7138–44

[271] Seo J W, Jun Y W, Park S W, Nah H, Moon T, Park B, Kim J G, Kim Y J and Cheon J 2007 Two-dimensional nanosheet crystals *Angew. Chem. Int. Ed.* **46** 8828–31

[272] Zheng G, Zhang Q, Cha J J, Yang Y, Li W, Seh Z W and Cui Y 2013 Amphiphilic surface modification of hollow carbon nanofibers for improved cycle life of lithium sulfur batteries *Nano Lett.* **13** 1265–70

[273] Sun K, Zhang Q, Bock D C, Tong X, Su D, Marschilok A C, Takeuchi K J, Takeuchi E S and Gan H 2017 Interaction of TiS2 and sulfur in Li-S battery system *J. Electrochem. Soc.* **164** A1291

[274] Zhao Y, Cai W, Fang Y, Ao H, Zhu Y and Qian Y 2019 Sulfur-deficient TiS_{2-x} for promoted polysulfide redox conversion in lithium-sulfur batteries *ChemElectroChem* **6** 2231–7

[275] Su Y S and Manthiram A 2014 Sulfur/lithium-insertion compound composite cathodes for Li–S batteries *J. Power Sources* **270** 101–5

[276] Pan S, Yin Z, Cheng Q, Zhang G, Yu X, Pan Z, Rao H and Zhong X 2020 Bifunctional TiS_2/CNT as efficient polysulfide barrier to improve the performance of lithium–sulfur battery *J. Alloys Compd.* **832** 154947

[277] Ma L, Wei S, Zhuang H L, Hendrickson K E, Hennig R G and Archer L A 2015 Hybrid cathode architectures for lithium batteries based on TiS_2 and sulfur *J. Mater. Chem.* A **3** 19857–66

[278] Chung S H, Luo L and Manthiram A 2018 TiS_2–polysulfide hybrid cathode with high sulfur loading and low electrolyte consumption for lithium–sulfur batteries *ACS Energy Lett.* **3** 568–73

[279] Huang X *et al* 2019 Sandwich-like ultrathin TiS_2 nanosheets confined within N, S codoped porous carbon as an effective polysulfide promoter in lithium-sulfur batteries *Adv. Energy Mater.* **9** 1901872

[280] Ma Z, Li Z, Hu K, Liu D, Huo J and Wang S 2016 The enhancement of polysulfide absorbsion in LiS batteries by hierarchically porous CoS_2/carbon paper interlayer *J. Power Sources* **325** 71–8

[281] Luo S, Zheng C, Sun W, Wang Y, Ke J, Guo Q, Liu S, Hong X, Li Y and Xie W 2018 Multi-functional CoS_2-NC porous carbon composite derived from metal-organic frameworks for high performance lithium–sulfur batteries *Electrochim. Acta* **289** 94–103

[282] Wang H, Zhang W, Xu J and Guo Z 2018 Advances in polar materials for lithium–sulfur batteries *Adv. Funct. Mater.* **28** 1707520

[283] Hong X, Li S, Tang X, Sun Z and Li F 2018 Self-supporting porous CoS_2/rGO sulfur host prepared by bottom-up assembly for lithium–sulfur batteries *J. Alloys Compd.* **749** 586–93

[284] Zhang H, Zou M, Zhao W, Wang Y, Chen Y, Wu Y, Dai L and Cao A 2019 Highly dispersed catalytic Co_3S_4 among a hierarchical carbon nanostructure for high-rate and long-life lithium–sulfur batteries *ACS Nano* **13** 3982–91

[285] Chen T *et al* 2017 Metallic and polar Co_9S_8 inlaid carbon hollow nanopolyhedra as efficient polysulfide mediator for lithium–sulfur batteries *Nano Energy* **38** 239–48

[286] Seo S D, Park D, Park S and Kim D W 2019 'Brain-Coral-Like' mesoporous hollow CoS_2@N-doped graphitic carbon nanoshells as efficient sulfur reservoirs for lithium–sulfur batteries *Adv. Funct. Mater.* **29** 1903712

[287] Zhou J, Lin N, Cai W L, Guo C, Zhang K, Zhou J, Zhu Y and Qian Y 2016 Synthesis of S/CoS2 nanoparticles-embedded N-doped carbon polyhedrons from polyhedrons ZIF-67 and their properties in lithium-sulfur batteries *Electrochim. Acta* **218** 243–51

[288] He J, Chen Y and Manthiram A 2018 MOF-derived cobalt sulfide grown on 3D graphene foam as an efficient sulfur host for long-life lithium–sulfur batteries *iScience* **4** 36–43

[289] Wang Q, Zou R, Xia W, Ma J, Qiu B, Mahmood A, Zhao R, Yang Y, Xia D and Xu Q 2015 Facile synthesis of ultrasmall CoS_2 nanoparticles within thin N-doped porous carbon shell for high performance lithium-ion batteries *Small* **11** 2511–7

[290] Xu H and Manthiram A 2017 Hollow cobalt sulfide polyhedra-enabled long-life, high areal-capacity lithium–sulfur batteries *Nano Energy* **33** 124–9

[291] Pu J, Shen Z, Zheng J, Wu W, Zhu C, Zhou Q, Zhang H and Pan F 2017 Multifunctional Co_3S_4@sulfur nanotubes for enhanced lithium–sulfur battery performance *Nano Energy* **37** 7–14

[292] Li X, Zhao K, Zhang L, Ding Z and Hu K 2017 MoS_2-decorated coaxial nanocable carbon aerogel composites as cathode materials for high performance lithium–sulfur batteries *J. Alloys Compd.* **692** 40–8

[293] Lin H, Yang L, Jiang X, Li G, Zhang T, Yao Q, Zheng G W and Lee J Y 2017 Electrocatalysis of polysulfide conversion by sulfur-deficient MoS_2 nanoflakes for lithium–sulfur batteries *Energy Environ. Sci.* **10** 1476–86

[294] Yan L, Luo N, Kong W, Luo S, Wu H, Jiang K, Li Q, Fan S, Duan W and Wang J 2018 Enhanced performance of lithium–sulfur batteries with an ultrathin and lightweight MoS_2/carbon nanotube interlayer *J. Power Sources* **389** 169–77

[295] Li Z, Deng S, Xu R, Wei L, Su X and adWu M 2017 Combination of nitrogen-doped graphene with MoS_2 nanoclusters for improved Li–S battery cathode: synthetic effect between 2D components *Electrochim. Acta* **252** 200–7

[296] Zhao Y, Zhuang Q, Li W, Peng H, Li G and Zhang Z 2019 Encapsulation of few-layer MoS_2 in the pores of mesoporous carbon hollow spheres for lithium–sulfur batteries *Nanomaterials* **9** 1247

[297] Zhang C L, Jiang Z H, Lu B R, Liu J T, Cao F H, Li H, Yu Z L and Yu S H 2019 MoS_2 nanoplates assembled on electrospun polyacrylonitrile-metal organic framework-derived carbon fibers for lithium storage *Nano Energy* **61** 104–10

[298] Walle M D, Zeng K, Zhang M, Li Y and Liu Y N 2019 Flower-like molybdenum disulfide/carbon nanotubes composites for high sulfur utilization and high-performance lithium–sulfur battery cathodes *Appl. Surf. Sci.* **473** 540–7

[299] Wu J, You N, Li X, Zeng H, Li S, Xue Z, Ye Y and Xie X 2019 SiO_2@MoS_2 core–shell nanocomposite layers with high Li-ion diffusion as a triple polysulfide shield for high performance lithium–sulfur batteries *J. Mater. Chem.* A **7** 7644–53

[300] Park J S, Jung Y and Kim S 2018 In situ synthesis of graphene nanosheet/MoS_2 composite electrodes and their electrochemical performance for lithium secondary cells *J. Nanosci. Nanotechnol.* **18** 44–7

[301] You Y, Ye Y, Wei M, Sun W, Tang Q, Zhang J, Chen X, Li H and Xu J 2019 Three-dimensional MoS_2/rGO foams as efficient sulfur hosts for high-performance lithium–sulfur batteries *Chem. Eng. J.* **355** 671–8

[302] Hu L, Dai C, Lim J M, Chen Y, Lian X, Wang M, Li Y, Xiao P, Henkelman G and Xu M 2018 A highly efficient double-hierarchical sulfur host for advanced lithium–sulfur batteries *Chem. Sci.* **9** 666–75

[303] Sun F, Tang H, Zhang B, Li X, Yin C, Yue Z, Zhou L, Li Y and Shi J 2018 PEO-linked MoS_2-graphene nanocomposites with 2D polar–nonpolar amphoteric surfaces as sulfur hosts for high-performance Li–S batteries *ACS Sustain. Chem. Eng.* **6** 974–82

[304] Babu G, Masurkar N, Al Salem H and Arava L M R 2017 Transition metal dichalcogenide atomic layers for lithium polysulfides electrocatalysis *J. Am. Chem. Soc.* **139** 171–8

[305] Lei T, Chen W, Huang J, Yan C, Sun H, Wang C, Zhang W, Li Y and Xiong J 2017 Multifunctional layered WS2 nanosheets for enhancing the performance of lithium–sulfur batteries *Adv. Energy Mater.* **7** 1601843

[306] Ye C, Zhang L, Guo C, Li D, Vasileff A, Wang H and Qiao S Z 2017 A 3D hybrid of chemically coupled nickel sulfide and hollow carbon spheres for high performance lithium–sulfur batteries *Adv. Funct. Mater.* **27** 1702524

[307] Zhang S S and Tran D T 2016 Pyrite FeS_2 as an efficient adsorbent of lithium polysulphide for improved lithium–sulphur batteries *J. Mater. Chem.* A **4** 4371–4

[308] Cheng Z, Xiao Z, Pan H, Wang S and Wang R 2018 Elastic sandwich-type rGO–VS_2/S composites with high tap density: structural and chemical cooperativity enabling lithium–sulfur batteries with high energy density *Adv. Energy Mater.* **8** 1702337

[309] Dai C *et al* 2018 Chinese knot-like electrode design for advanced Li–S batteries *Nano Energy* **53** 354–61
[310] Xiao Z, Yang Z, Zhang L, Pan H and Wang R 2017 Sandwich-type NbS_2@S@ I-doped graphene for high-sulfur-loaded, ultrahigh-rate, and long-life lithium–sulfur batteries *ACS Nano* **11** 8488–98
[311] Liu J D, Zheng X S, Shi Z F and Zhang S Q 2014 Sulfur/mesoporous carbon composites combined with γ-MnS as cathode materials for lithium/sulfur batteries *Ionics* **20** 659–64
[312] Kim J G, Noh Y and Kim Y 2022 Highly reversible lithium–sulfur batteries with nitrogen-doped carbon encapsulated sulfur cathode and nitrogen-doped carbon-coated ZnS anode *Chem. Eng. J.* **435** 131339
[313] Li X, Hu K, Tang R, Zhao K and Ding Y 2016 CuS quantum dot modified carbon aerogel as an immobilizer for lithium polysulfides for high-performance lithium–sulfur batteries *RSC Adv.* **6** 71319–27
[314] Wang H, Li J, Li K, Lin Y, Chen J, Gao L, Nicolosi V, Xiao X and Lee J M 2021 Transition metal nitrides for electrochemical energy applications *Chem. Soc. Rev.* **50** 1354–90
[315] Gao B, Li X, Ding K, Huang C, Li Q, Chu P K and Huo K 2019 Recent progress in nanostructured transition metal nitrides for advanced electrochemical energy storage *J. Mater. Chem.* A **7** 14–37
[316] Yuan H *et al* 2017 Efficient activation of Li_2S by transition metal phosphides nanoparticles for highly stable lithium–sulfur batteries *ACS Energy Lett.* **2** 1711–9
[317] Shi Z, Feng W and Wang X 2022 Ni_2P/carbon nanotube nanocomposite as host material for high performance lithium–sulfur battery cathode *Chem. Phys.* **553** 111383
[318] Hart J L, Hantanasirisakul K, Lang A C, Anasori B, Gogotsi Y and Taheri M L 2018 Direct correlation of mxene surface chemistry and electronic properties *Microsc. Microanal.* **24** 1606–7
[319] Xiao Z, Li Z, Meng X and Wang R 2019 MXene-engineered lithium–sulfur batteries *J. Mater. Chem.* A **7** 22730–43
[320] Tang X, Guo X, Wu W and Wang G 2018 2D metal carbides and nitrides (MXenes) as high-performance electrode materials for Lithium-based batteries *Adv. Energy Mater.* **8** 1801897
[321] Pang J, Mendes R G, Bachmatiuk A, Zhao L, Ta H Q, Gemming T, Liu H, Liu Z and Rummeli M H 2019 Applications of 2D MXenes in energy conversion and storage systems *Chem. Soc. Rev.* **48** 72–133
[322] Dong Y, Zheng S, Qin J, Zhao X, Shi H, Wang X, Chen J and Wu Z S 2018 All-MXene-based integrated electrode constructed by Ti_3C_2 nanoribbon framework host and nanosheet interlayer for high-energy-density Li–S batteries *Acs Nano* **12** 2381–8
[323] Liang X, Garsuch A and Nazar L F 2015 Sulfur cathodes based on conductive MXene nanosheets for high-performance lithium–sulfur batteries *Angew. Chem.* **127** 3979–83
[324] Lv L P, Guo C F, Sun W and Wang Y 2019 Strong surface-bound sulfur in carbon nanotube bridged hierarchical Mo_2C-based MXene nanosheets for lithium–sulfur batteries *Small* **15** 1804338
[325] Zheng Y, Zheng S, Xue H and Pang H 2019 Metal–organic frameworks for lithium–sulfur batteries *J. Mater. Chem.* A **7** 3469–91
[326] Xu J, Lawson T, Fan H, Su D and Wang G 2018 Updated metal compounds (MOFs, –S, –OH, –N, –C) used as cathode materials for lithium–sulfur batteries *Adv. Energy Mater.* **8** 1702607
[327] Zhong Y, Xu X, Liu Y, Wang W and Shao Z 2018 Recent progress in metal–organic frameworks for lithium–sulfur batteries *Polyhedron* **155** 464–84

www.ingramcontent.com/pod-product-compliance
Ingram Content Group UK Ltd.
Pitfield, Milton Keynes, MK11 3LW, UK
UKHW051852100926
8556IPUK00005B/589